# Neuroelectricity
## The Past, Present, and Future of Brain Stimulation

Kyle Kirkland

# Contents

# Preface

How does the brain work? How do our conscious thoughts and feelings arise from a lump of organic material bathed in brine? What makes people think and behave the way that they do?

I wish I could answer those questions. I also wish I could express optimism that somebody will soon find the answers, but this is highly unlikely. The brain is too complex, and although it is this complexity that gives us the ability to ponder such deep questions, it also poses a great challenge for the relative simplicity of our science. Brains were made for survival and the perpetuation of the species, not for solving scientific quandaries. The best we can probably achieve in the near future is a clearing of the mist here and there, a glimpse of some deeper truth.

Neuroscience is the branch of science dealing with the brain, and neuroscientists have fashioned many tools and techniques over the years to help them in their studies. After discovering the importance of electrical activity, neuroscientists immediately began plying the brain with electric currents and observing what happens. Often something interesting happens, and people have gotten many fascinating glimpses of the functions of the brain with these stimulation experiments. That's what this book is about.

Researchers have also discovered that electrical stimulation has the power to change the brain. Sometimes it can be used to alleviate symptoms of neurological or psychiatric disorders. In other times it can be used to manipulate behavior and emotions. These technological applications—and their potential for good or evil—is also what the book is about.

I'm not going to define precisely what I mean by electrical stimulation of the brain. You can twist yourself into knots trying to be rigorous about definitions in neuroscience, even about things that appear simple. Let's just say that the topics in this book involve encouraging electric charge to start flowing somewhere in the nervous system by artificial means.

This book isn't a medical treatise and I don't offer medical advice. The intended audience is anyone who is interested in brain stimulation. But even though I'm not specifically writing for patients who are presently undergoing or contemplating treatments involving brain stimulation, I hope the book can be useful to these people and help them understand the concepts related to these treatments.

Another caveat: I am an opinionated person and I'm not shy about it. That's what you get, or what you're supposed to get, when you buy a book—the author's point of view. I have experience in this field of research and I was at one time a neuroscientist, earning a Ph.D. and working in various laboratories. I cite data to support my conclusions, though I also try to point out any contrary evidence that people holding a different opinion would find significant. This sometimes makes me seem ambivalent, and so I am on a lot of things—rarely do I find anything that is unassailably good or thoroughly bad. I also don't shy away from offering my subjective viewpoints, based more on my feelings than on facts. That, too, is an author's prerogative. But I am fully aware that reasonable and intelligent people hold beliefs completely opposite of my own. I have the good fortune of being opinionated yet able to avoid the delusion of infallibility.

Journalists are generally careful to avoid jargon, but I'm also a scientist, or used to be, and so I'm more comfortable with scientific terminology than other writers. My scientific background can be an advantage for me as a writer because I have a lot of experience in the field, but it can also be a problem because I'll write phrases like thalamocortical rhythms or neural synchrony without a second thought, forgetting that most people have little idea what these terms mean. I've tried to avoid jargon in this book, and I've provided a glossary in the back for those terms I couldn't help using.

Besides my neuroscience doctorate, my qualifications for writing this book include research experience in laboratories that did a lot of stimulation experiments, though in animals rather than humans. I also have a good bit of knowledge about electrical equipment: before I went to college I did a hitch in the Air Force as an instrumentation technician, so I know something about electricity and getting shocked (never in the brain, though, and rarely intentionally). As for the futuristic aspects of the book, my experience includes writing a number of science fiction stories and spending a lot of time imagining where the next advances in science and technology may take us.

I could have written a book ten times this size and still had material left over, but the book is big enough as it is. Selectivity was unfortunate but necessary.

Authors often end their prefaces by acknowledging the contributions of other people and then go on to say that no book is ever really the work of one person. I disagree.

# 1
## Electricity and the Brain

As a first-year neuroscience graduate student, I had the opportunity to hold a human brain in my hands. It was a real human brain—but not, of course, a living one. Wet, brownish, wrinkled, smelling of formaldehyde, it had long since ceased to function as someone's mind.

It gave me mixed emotions. Sadness, because it was now an inert, lifeless thing; amazement, because of what it once had been.

I stared at it, mesmerized, for a considerable period of time. Which wasn't what I was supposed to do. This was an anatomy class and it was dissection time. The other students were busy. But it was hard for me to do anything but hold the brain at arm's length and look at it.

"It doesn't bite," prodded the professor.

True enough. And there was no little person, no "homunculus" inside it either. It was a chunk of tissue, perfused with a solution that had hardened it to the consistency of cheese.

But wasn't it wrong, I wondered—was it even almost sacrilege—to defile this tomb? That's how I regarded it: a tomb. The object I held in my hands, an object which profoundly amazed and puzzled me, was the tomb of a mind.

I knew nothing of the person whose thoughts and feelings had once swirled around inside this brain, back when it was a living, pulsating group of cells called neurons connected together by little junctions called synapses. Gender, age, race, physical appearance—I couldn't be sure of anything about the person who had once lived in this brain. All I knew was that for some reason the person had seen fit to donate, in a charitable bequest, his or her body to science.

Finally I overcame my inhibition. Like the other students, I cut, probed, pinched, pulled, and tore. Scattered parts of the brain began to fill the metal pan on which I worked. I identified this and that structure. And when it was over the professor gathered together all the pans, covered them, and locked them up in a cabinet. The brain parts would be buried, we were told. But before they were sent away, the professor said that we would be allowed another look, if we wished. The lab would be open that evening for any student who wanted to spend more time studying the brains.

I came back for another look. No one else did, but I sat there, alone, for hours. I arranged and rearranged the bits and pieces of tissue, and I tried to imagine how they had once looked. Try as I might, it was nearly impossible for me to understand how minds had once inhabited them.

I certainly wasn't alone in thinking this way. Many people, now and in the past, have had a hard time believing that the brain is solely responsible for the mind. Neuroscientists speak of it as a fact, yet it's almost inconceivable. How ridiculous it is to consider the dynamic cauldron of emotions and thoughts that makes up every living human being, and claim that the source of all of it lies within this little three pounds of unremarkable tissue—proteins, lipids, carbohydrates, water, and very little else.

The famous ancient Greek philosopher Aristotle didn't believe it. He considered the source of intelligence to be the heart instead of the brain. Not that that's really any more philosophically or physiologically plausible, but at least he chose an obviously important organ for the job. The brain, according to this noted philosopher of the ancient world, had other functions, like cooling the blood.

Perhaps Aristotle got this idea by observing headless chickens, which occasionally run around the barnyard after decapitation. They're not the only ones capable of such feats: frogs are such "spinal animals" that some 19th century scientists, after studying this species, were convinced that the spinal cord alone was sufficient. Who needs a brain?

Humans do, as unfortunate victims of brain injury prove. Even back in Aristotle's day— and particularly in Roman times, when people were so fond of gladiator combat—people started to recognize that damage to the brain leads to changes in behavior, personality, and thinking patterns. What these early scientists and physicians didn't realize is that the electric fish that they found in the Mediterranean Sea—and greatly marveled over—had a great deal in common with the way the brain works.

The brain is critical for thinking. And yet...you can take the brain apart, cut it up and grind it up and put it under a microscope and identify this molecule and that molecule and record tiny bioelectric currents and capture images of metabolic processes and trace neuron projections called axons that snake through tissue and make synapses with other neurons...and when you've done as much as you possibly can and sit back and think about what you've seen, it's *still* nearly impossible to imagine that the mind actually lives there.

But you don't have to take it apart to study it. Neuroscientists have found ways to examine the brain while it's still in use, locked up tight in the skull. Scientists can correlate the activity of the brain with a person's behavior and thoughts and emotions. If the person moves this way, a certain part of the brain becomes more active; if he or she feels a particular emotion, another part of the brain gets activated. This doesn't necessarily mean the brain actually causes these movements and feelings, but it does mean the brain is in some way intimately related to them.

Even so, it's still not quite convincing enough. I guess that's because it's so difficult to believe that the brain actually creates the mind. So why do a lot of people, including virtually all neuroscientists and even including my own skeptical self, believe it?

The proof turns out to be in an electrode—an electrical conductor that can not only measure current but can also inject it into the brain. Sending current through the brain turns out to be quite stimulating.

* * *

Brad[1] is twenty-six years old. Tall and handsome, he has an athletic build, though it's clear he's been a bit more idle these days than in the past: a small but detectable paunch bulges out from his blue New York Giants T-shirt. He jokes with the nurse who's carefully unwrapping the bandage covering his head. "They should hire me for one of those alien movies," he says.

About a year ago Brad was diagnosed with epilepsy. A little less than two years before that he was involved in an automobile accident and suffered a serious concussion. Brad, young and healthy, recovered quickly, and for a while it looked like there would be no lasting injuries. Then the seizures started.

They often take a while to appear. No one knows exactly why, but every once in a while epilepsy will develop months, even years after an incidence of head trauma. Fortunately it's not an inevitable consequence; but for those who are unlucky, that matters little.

Brad's seizures are a particularly tough variety to control. The seizures seem to start in the temporal lobe, the part of the brain that's at the side of the head. But the seizures spread to other parts of his brain, including, and most conspicuously, the part of the brain that controls his muscles. During a seizure Brad's muscles go haywire, he often becomes incontinent, he loses consciousness, and he falls down. Such episodes, sometimes called convulsions, happen several times a day, at unpredictable times. Often he does get a brief warning—an aura, as it's called. How an aura is experienced depends on the patient; Brad describes his as sort of a feeling of déjà vu. The aura tells him he's probably about to have another seizure in the next moment or two. Actually, it tells him he's probably already having one. But since Brad gets no warning well in advance of a seizure, he doesn't know when one will strike. His activities are thus severely curtailed: he can't drive, he can't even ride a bicycle, he can't do anything in which a sudden interruption would prove dangerous to himself or anyone nearby.

Epilepsy is still poorly understood, even though a number of drugs have been developed which have helped millions of people with the disorder. About 1 percent or so of the population will at some time struggle with one kind of seizure disorder or another. Fortunately, most will be helped by medication—at least one of the many antiepileptic drugs will work (no one knows exactly how), so that the frequency of seizures is reduced to a tolerable level and intensity, and in some cases eliminated entirely. But for about 25 percent of epilepsy patients, either none of the drugs work or they don't work for very long. Brad is one of the 25 percent.

Neurologists know a lot more about seizures than they did in the days when they used terms like "grand mal" (meaning a seizure in which the patient has convulsive movements) and "petit mal" (what today would be called an absence seizure, in which the patient pauses and stares into space for a moment or two). Many different types of seizure have been identified, some of which are genetic and some of which are acquired from disease or injury. Seizures happen when part or all of the brain becomes uncontrollably activated—neurons began to discharge in perfect synchrony—and neurologists have developed ways of recording or imaging these events. If only part of the brain is involved, the patient may feel funny sensations or some of their muscles may twitch. Devices that record or image brain activity have been of great use. For example, the old-fashioned but still widely used electroencephalogram (EEG) uses electrodes pasted to the scalp to detect current leaking across the skull.[2] A more modern example is positron emission tomography (PET), which images brain metabolism that is correlated with the electrical activity of neurons.[3]

Sometimes it's only a tiny part of the brain that seems to cause the problem. This troublesome little area might be highly excitable, for example, and when it goes off it becomes an instigator—a lead domino that falls and may cause the rest of the brain to fall with it. The result is a succession of neural networks which stop what they're doing and start firing away, rapidly and synchronously.[4]

Neurologists have determined that this is what happens during one of Brad's seizures. It was an important discovery because knowing where the seizures begin may make it possible to treat Brad's heretofore untreatable epilepsy. But the doctors had to do something rather drastic in order to make this discovery.

The nurse finishes unwrapping Brad's bandage. He's got

wires coming out of his skull, and a technician begins to hook them up to a machine.

Neurons in the brain generate tiny electric currents, by which they produce messages that are passed from one neuron to another. Like all other current, neural electricity can be measured and recorded. It can even be detected from the surface of the head, which is commonly how the EEG is recorded. Sensitive electrodes are pasted onto the scalp, or are weaved into a snugly fitting skull cap, and produce a signal proportional to the current that's leaking from the neurons in the brain.

But the surface (i.e., scalp) EEG is a very noisy and difficult-to-interpret measurement. The small neural currents are greatly reduced and strongly filtered by the skull and scalp. What's worse, the muscles of the scalp and face produce electric currents as well, and since they're closer to the electrodes than the brain is, they can contaminate and overwhelm the minuscule signals from neurons. Another problem with surface EEG is that even under ideal conditions, it's problematic or impossible to pinpoint the exact area of the brain from which the signals are coming from.

Imaging techniques such as PET and functional magnetic resonance imaging (fMRI) are much superior to scalp EEG for locating which parts of the brain are active.[5] PET and fMRI are two methods of getting a three-dimensional picture of neural activity. These imaging techniques give you a 3D view with good spatial resolution—for the best fMRI machines the resolution is a millimeter or so. This means you can distinguish between two points even if they are separated by only a millimeter, which is about 0.04 inches. That's pretty good resolution, even taking into consideration that the brain is a small organ, averaging about 1500 cubic centimeters (91 cubic inches) in volume, and crucial components can occupy an astonishingly small part of the total.[6] However, imaging isn't as useful as you might think. These devices may have good spatial resolution but they generally have poor temporal resolution—you may be able to determine *where* things are happening but you can't figure out exactly *when* they happen. Another big problem with imaging a seizure is that you don't know at what time the next one will occur, and unfortunately because of safety and other issues, you can't image the patient over a long period of time.

But physicians must determine exactly where the seizure begins, because the treatment that Brad will receive involves excising the diseased tissue. Cutting out a part of the brain can have dire consequences; obviously you want to remove as small a part as possible. But that means you have to know precisely where to cut.

The best way that neurologists can get this information is to insert thin electrodes into the brain. If the electrodes are tiny and are carefully placed, they do little damage. Although the procedure is invasive and thus undesirable, signals can be recorded right where the action is happening—in the brain itself. This gives doctors an accurate picture of the brain's activity, provided enough electrodes are used.

But the most fascinating thing about putting electrodes into the brain is not what happens when they pick up the small currents generated by neurons. The most fascinating thing is what happens when these electrodes produce some current of their own.

Electrical stimulation of the brain can have all kinds of effects, depending on where the electrode is and how much current is injected. Before neurosurgeons remove the epileptic portion of Brad's brain, they will carefully map the location and its surrounding area during the operation by applying small electric currents. By doing this they will get some idea of the tissue's function. For example, stimulating a certain part of the cerebral cortex—which is located on the brain's surface—will cause a certain muscle or set of muscles to contract. Stimulating in another place may cause a sensation to be felt—a touch on the shoulder, perhaps. So this, then, is the final piece of evidence proving that electrical activity of the brain is responsible for sensation and behavior. Not only are the currents recorded from the brain correlated with behaviors and perceptions, but the same behaviors and perceptions can be initiated with currents injected by electrodes. Electricity in the brain creates the mind.

In operations like the one performed on Brad, surgeons will carefully avoid removing certain parts of the brain. They won't cut out tissue that's vital for speech, for instance, even if it means leaving in a small portion of epileptic tissue.

When they inserted the electrodes into Brad's brain, neurologists tested them by passing small currents. When a certain part of Brad's brain is stimulated, he reports feeling "nostalgic." It seems to be a vague, indistinct sort of feeling, yet Brad consistently mentions it when that specific electrode is stimulated. It has to be a real effect, because neurologists can switch on the electrode at any random instant, and Brad won't know when they've done it. He has no way of knowing when the electrode was turned on because he can't feel the current itself; he can only know of its effects on the surrounding brain tissue.

With surgery Brad has a good chance of greatly reducing the number of seizures he suffers. Maybe he'll even be free of them entirely.

These procedures clearly have benefits. Many people, however, are uneasy. In a way, so is Brad. His "aliens" remark may have been just a joke, but more than likely it went a little deeper than that.

No one objects when surgeons implant an electrical device in the chest. Pacemakers for the heart are a wonderful technology. What's more, they are socially acceptable, and have been used ever since they were first developed in the 1950s. But the brain is different.

Science fiction writers have devoted considerable ingenuity into developing various scenarios involving electrical manipulation of the brain. They've written stories of "wireheads" who are helplessly addicted to electrical stimulation of the brain; stories about thought control, where minds are manipulated by ruthless dictators; stories about memory enhancers or erasers, instant learning, and the making (or unmaking) of geniuses — usually with less than ideal consequences. Few of these stories give readers very much reason to be optimistic about this sort of technology.

And the stigma remains. Heart, yes; brain, no. The brain is not a good place for electrodes.

Actually, this is quite true—but no part of the body is a hospitable place for little slivers of metal. It takes a bit of work to fashion workable electrodes and to put them in the right spot. But once this is done, it can—and does—make the lives of people like Brad a lot better.

The stigmas as well as the modern techniques are of course the result of a lot of history. It's been a bumpy ride. Science fiction writers might be surprised to learn that a lot of horrifying experiments they've dreamed up have, to a limited extent, already occurred.

Like every other scientific development, brain stimulation has brought about some good things and some bad things. More importantly, it has the potential to bring about a great deal more of each. Its future role is uncertain and debatable. But unquestionably it has also given us insight into the most amazing object in the universe.

* * *

One of the first people to stimulate the human brain with electricity was Italian scientist Alessandro Volta. Ancient civilizations had discovered and sometimes therapeutically used electric fish, applying them to various parts of the body (including the head), but it was Volta who brought about the means of producing a steady and controlled current. (It's from Volta's name that the unit of electric potential, the volt, is derived.) Around 1800 this venturesome scientist placed one electrode into each ear and hooked them up to his new invention, the Voltaic pile, which these days we'd call a battery. Volta reported hearing a strange noise and experienced a *"secousse dans la tête* [jolt in the head]."[7] Wisely, he decided not to repeat the experiment very many times.

People had been fooling around with electricity long before they had any idea what it was or what caused it. But they did notice one thing, and it was actually Volta's contemporary and fellow Italian, Luigi Galvani, who publicized it, though others had observed it earlier.[8] In one series of experiments in the latter part of the 18th century, Galvani used electricity generated by dissimilar metals (which produce a small voltage when in contact with each other) to demonstrate the intimate connection between electricity and nerve. He found that when he sent a current into the nerves or muscle of frog legs, they twitched. Galvani managed to convince himself and a number of other people that "animal electricity" was an integral part of the body.

In fact, it is. But not quite in the way that Galvani figured. Muscles are electrically active tissues, as are the cells in the brain called neurons. (Nerves are composed of long processes of neurons called axons). Like neurons, muscle cells have large numbers of proteins called ion channels spanning the cell's membrane. The membrane is a very thin layer of lipids and proteins that encloses the cell and maintains a sort of traffic control. It provides the necessary separation between the cell's interior and the exterior environment, and controls the flow of substances between the two. One of these substances happens to be ions, which are electrically charged particles in solution. If you dissolve salt (sodium chloride) in water, for instance, you get ions of sodium, having a positive charge, and ions of chloride, having a negative charge. Our tissues contain a lot of different ions. It's the job of ion channels to control the flow of ions that cross the membrane. This flow of charge constitutes an electric current, the same as if it was flowing through a wire.

Certain ion channels in brain and muscle (including the heart) have an amazing property: they're sensitive to voltage. Across the membrane of all cells there's a small voltage, which occurs naturally because of certain electrochemical properties. But in brain and muscle cells this voltage can change. Some of the more important ion channels work to produce a critical event called an action potential. Generated by the opening and closing of certain ion channels, action potentials are transient changes in the membrane potential, which rises and then falls. The whole thing lasts about a millisecond or two. Action potentials underlie all muscular contraction, including heartbeats. Action potentials also occur in neurons as well as muscle cells, but for a different purpose; in muscles an action potential initiates contraction, but in neurons an action potential is used to convey information.

In the skeletal muscles, action potentials are generally evoked by messages from nerves. Nerves are those white stringy things that Descartes believed were full of spirits but which are actually the business end of neurons. When a neuron fires an action potential, the impulse travels down the axon and causes the release of a tiny spurt of chemicals called neurotransmitters. The neurotransmitters produce a small rise in the electrical potential of the recipient cell, which can be another neuron but can also be a muscle cell. In muscle, the rise in potential caused by the neuro-transmitters may invoke a full-blown action potential (thanks to the ion channels which are sensitive to voltage) and, consequently, a contraction. Usually the decision to cause a muscle to contract or not comes from neural computations done in the spinal cord or the brain, the results of which are conveyed to the muscles by way of nerves. However, when Galvani stimulated the nerve or a muscle of a frog leg, he bypassed all of that top-level stuff and merely kicked off an action potential via the current from his electrode. He believed that his experiment mimicked the way that movement naturally happened, and so it did, in a way. But the real boss of the show, the brain, was being left out.

People began to wonder about that. The brain was the boss, as scientists had already discovered. Did it work by electricity as well? That's what motivated Volta to perform his experiment, from which he was fortunate enough to survive. But he was no wiser, alas.

What would happen if you injected electric current directly into the brain? Physiologists studying the brain of animals began performing such experiments in the 19th century. And then came a physician named Roberts Bartholow who was bold enough to do it in a human.

## 2

## The First Human Brain Stimulation: The Tragedy of Mary Rafferty

Roberts Bartholow was a man of unimposing stature and a notable lack of congeniality. Born in New Windsor, Maryland in 1831, he was diligent in school and earned a medical degree from the University of Maryland in 1852. His stern disposition and controlled, disciplined aggression helped him endure the challenging years he spent as an army surgeon in the Wild West of the late 1850s and early 60s. A sense of loyalty impelled him to return to Baltimore during the American Civil War, and subsequently he served at various hospitals in the Union army. In 1864, after a career spanning seven years, he left the military and he moved his family to Cincinnati, Ohio, where he established a medical practice.

Bartholow's success came quick. He built a thriving practice and achieved an excellent reputation in the field of medicine.[1] Cincinnati possessed facilities in the medical arts of no small importance to the burgeoning Midwest, and Bartholow fit in well. He obtained important posts at local hospitals and colleges, such as the Medical College of Ohio, and became a leader in the incipient field of medical research. His pioneering spirit was evident throughout. He was also a prolific author. He founded a medical journal called *The Clinic*; his editorials were vibrant, enthusiastic, and at times quite caustic, for the progressive Bartholow pulled few punches in his criticism of his contemporaries, and his stinging commentaries rankled more than a few of his colleagues and fellow physicians. Bartholow may have left the army but he had taken his fighting spirit with him.

Ten years after coming to Cincinnati, Bartholow became the first surgeon to report an invasive electrical stimulation of a human brain—a stimulation in which electrodes were inserted into the brain rather than attached to the scalp—and became embroiled in a controversy that soon threatened to engulf him.

Bartholow was a firm believer in the old dictum of *carpe diem* (seize the day), and perhaps it was this, more than anything else, that got him into trouble. But it was also this conviction that had served him so well in the years preceding 1874. It was a good thing to have in those days, an era of great change in surgical procedures. And most of the change was for the better.

In the early years of surgery, operations were horrific and gory. Morbidity was common and accepted as such. Surgery was not neat, precise, or even aseptic, and the only anesthesia was an intoxicating dose of alcohol or a severe blow to the head. The principal skills of a surgeon were his speed and ability to hit a moving target.

Gradually, in the latter half of the 19th century, all of that changed. Reasonably safe agents for anesthesia were discovered, and the incidence of infection was much reduced by the observance of simple but crucial methods like washing and sterilizing. Going "under the knife" slowly lost its close kinship with torture or a death sentence.

Other changes were more controversial. Darwin's theory of evolution, introduced in 1859, implied that animals and humans were related in physiologically important ways. If that's the case, then it made a lot sense to study animal biology in order to learn something about humans. This was highly relevant to the emerging field of medicine, since it suggested that you could test drugs and procedures in animals first. Presumably this made procedures safer for humans, if the knowledge gathered from animals was applicable to our species, which evolutionary theory claimed was true. Of course it meant that animals became "guinea pigs," and experimental treatments are often dangerous, painful, and crude. Proponents of animal experimentation pointed out that it was better for an animal to suffer than a person; opponents—of which there were many—could just as easily point out that if evolutionary theory was true then it meant animals *would* suffer, since they are like us in many important respects and this would presumably be the case for feeling pain as well. Bartholow was an enthusiastic supporter of the use of animals in research, and seemed very little moved by the ire of the antivivisectionists who vehemently opposed it.

All went quite well with Bartholow's career until 1874. In late January of that year a patient suffering from a severe scalp ulcer was admitted to Good Samaritan Hospital in Cincinnati. Her name was Mary Rafferty, a thirty-year-old "domestic" (household servant). Her wound was hideous, described as purulent (which meant it was secreting pus) and cancerous. Over the previous year the ulcer had increased in size, eventually eroding the skull to the point where several square inches of Mary's brain was exposed.

It seems that when Mary was a baby she had fallen into a fire, and a portion of her scalp had been severely burned. The hair never grew back. Subsequently Mary had gotten into the habit of wearing a wig, and it was this, she told her doctors, which led to her ulcer: a constant irritation from a piece of whalebone in the wig was thought to be the instigator.

However it happened, Mary was clearly in trouble. But she maintained a cheerful disposition, as noted by one of the physicians who interviewed her. He noted something else, too: her intelligence was well below normal, although the degree of her deficit was not recorded. In Bartholow's words she was "rather feeble-minded."[2]

Over the next few weeks the physicians did all they could for Mary. They drained the pus from the rapidly growing abscesses, but infection was inexorably taking its toll. The doctors fought a valiant but losing battle. The prognosis was bleak. Mary didn't have much longer to live.

One of the physicians interested in the case was Roberts Bartholow. In February he conceived of a bold idea. *Carpe diem* was undoubtedly uppermost in his mind.

What would happen if Mary's brain was stimulated with electricity? It would be easy to do, since part of her cranium had eroded completely away and a large portion of the surface of her brain—the cerebral cortex—was already conveniently accessible.

The idea of electrically stimulating the brain had plenty of precedence, as any well-read medical man of the time could have told you. One especially noteworthy series of experiments had been done but a few years earlier, mostly in dogs, by a pair of German researchers. The unexpected results had rapidly become famous. Articles on and references to this piece of research were ubiquitous, scattered in journals far and wide. A British physician had subsequently done several equally well known and much discussed experiments in monkeys and a number of other species. But no one had yet done the experiment in a *human*.

Perhaps Bartholow was also motivated by a similar case that happened many years earlier, involving not the brain but rather the stomach. In 1822 William Beaumont, an American physician, treated an 18-year-old man by the name of Alexis St. Martin, a victim of a gunshot wound in the abdomen. St. Martin's wound left a hole that refused to heal completely, and Beaumont could observe the young man's stomach in action. Since very little was known about digestion at the time, Beaumont realized that he had excellent opportunity to make an important set of observations and experiments. He did just that, spending years studying the process of digestion, as revealed in St. Martin's abdominal cavity. Taking full advantage of his lucky—and St. Martin's unlucky— circumstance, Beaumont discovered many important facts in his measurements and observations, and his work is cited today as instrumental in establishing the important fact that digestion is a chemical process.

Although it's not clear that this played a role in Bartholow's thinking, it *is* clear that he knew about the animal experiments. He specifically mentioned them in his published report.

The German scientists, Gustav Fritsch and Eduard Hitzig, had begun to publish their results four years earlier, in 1870. Previous to this, other researchers had repeatedly exposed the cerebral cortex of animals and stimulated it with electricity, only to find little result. But Fritsch and Hitzig were not deterred. Fritsch became interested in this sort of experiment because of his war experiences: while dressing head wounds of soldiers, he had noticed that irritation of the brain inexplicably caused reflexive movements in the *opposite* side of the body. Hitzig had supposedly become interested in the subject because he'd passed electric current through the heads of humans and observed that the process seemed to evoke eye movements.[3]

But the dogma that the cerebral cortex was electrically inert was strong at the time that Fritsch and Hitzig began their experiments. In addition to all of the earlier, failed experiments—and perhaps, to a certain extent, because of them—the cerebral cortex was taken to be the seat of intelligence by many, and, as such, devoid of the primitive feature of electrical irritability. To escape embarrassing comment from their academic colleagues Fritsch and Hitzig did their work outside of the lab. They used a dressing table in Hitzig's house in Berlin, a move that went over none too well with Frau Hitzig.

Back in those days you were forced to make your own electricity. There were no electrical outlets serviced by the local power company. With the proper equipment you could, however, produce whatever type of current you wanted to make from the following choices: "franklinic" electricity, also known as static electricity and named after Benjamin Franklin, the American statesman and scientist of kite-in-the-thunderstorm fame; "galvanic" current (named in honor of Galvani) which was what we would call direct current, like that which comes from a battery; and "faradic" current, named after the British scientist Michael Faraday, which consisted of pulses of electricity generated by an induction coil, similar to AC (alternating current).

Perhaps because of the care with which they did the experiments, Fritsch and Hitzig were successful where their predecessors had failed. They exposed the cerebral cortex of dogs and a few other types of animal and, by stimulating with weak galvanic currents, they managed to elicit a number of responses. Chief among these were twitches or contractions of muscles in the opposite side of the body. The brain of all mammals, including humans, is bilateral, composed of two roughly identical halves or cerebral hemispheres. Covering the surface of these hemispheres is a thin layer (actually several layers) of neurons—the cerebral cortex, or just cortex for short.[4] Fritsch and Hitzig found that current injected into one area of the anterior (front) part of the cortex on the brain's left hemisphere evoked twitches from the right forepaw. A strong stimulus would initiate a convulsion. Even more astonishing to scientists at the time, the German researchers discovered that there was only a part of the cortex which was related to movement, and it was in approximately the same spot in every animal. Fritsch and Hitzig had made the monumental discovery of cerebral localization, one of the main principles of modern neuroscience.

Initially the general opinion went against Fritsch and Hitzig, though not because of their crude experiments, which despite the ammunition it gave antivivisectionists was fairly common among biologists at the time. Fritsch and Hitzig weren't believed simply because their findings were too amazing.

However, the initial skepticism of the research community grudgingly gave way to acceptance, as the work was repeated and extended. One of the main researchers to do this was British physician David Ferrier, a man whose fame and reputation would grow to the point where he came to be considered one of the founders of experimental neurology.

Ferrier was very much interested in epilepsy. He wanted to study convulsions in animals by electrically stimulating their brains. By doing these experiments he hoped to get a handle on what causes the disease in humans. Using monkeys, Ferrier managed to evoke seizures by electrical stimulation, as well as movements similar to those of chorea (a disease characterized by spasms which produce movements resembling dancing).

Ferrier was also quite curious about the effects of stimulation upon the mind as well as the brain, and, though he refused to speculate too deeply, he wondered what sort of effect it was having on the animals. He wrote: "There is reason to believe that when the different parts of the brain are stimulated, ideas are excited in the animals experimented upon, but it is difficult to say what the ideas are."[5]

Another interesting aspect of the work of the two Germans

and the Englishman was their acrimonious relationship with one other. You might think they would have been more collegial, since each one's work complemented the other, but this was not the case. They fought bitterly over how to interpret the details of the experiments and over who should receive credit for what. The dispute took on an ugly nationalistic flavor, with the German scientific community championing Fritsch and Hitzig and denigrating the work of Ferrier, and British researchers doing just the opposite.[6]

And so amidst this furor and scientific upheaval and nationalistic clamor, in walks an American, Roberts Bartholow, with, as Americans are wont to have, a really *big* idea. With perhaps a starry gleam in his eye and visions of fame dancing upon his head, Bartholow asked Mary Rafferty for permission to begin a series of experiments to stimulate her brain with electricity. Bartholow claimed that she cheerfully gave her consent.

Not only would Bartholow be using a live person in his experiment—and thus his work would be directly relevant to medicine—his subject, unlike Fritsch and Hitzig's dogs and Ferrier's monkeys, could *describe* what she felt when she was being stimulated.

But it was an experiment nonetheless. There was no indication that Bartholow believed it would do the slightest bit of good for the patient. Bartholow had tossed away his physician's hat and had become, for the moment, purely a researcher. His subject was a member of the species Homo sapiens.

Bartholow used faradic current. He was fortunate in that he had available all the equipment he would need, since every hospital, including the Good Samaritan in Cincinnati of 1874, contained plenty of electrical equipment. It was the heyday of electrotherapy, and any hospital without the means of providing electrotherapeutic treatment was woefully out of date indeed. The progressive Bartholow was quite familiar with its use.

In the first experiment Bartholow tested for painful reactions to a mechanical stimulus in his conscious, awake patient. He discovered that while there was pain associated with the tissue surrounding the brain, there was none when the probing was limited to the brain itself. At least that's what Mary indicated, and her lack of any physical reaction seemed to confirm it. As Bartholow wrote in his report, "No pain whatever was experienced in the brain-substance proper." It was only the first experiment but Bartholow had already made an important discovery: there are no pain receptors or nerves in the brain. Bartholow was encouraged to proceed further.

The next step was to send current through the meninges. The brain is surrounded by several layers of membranes, collectively known as the meninges. They play an important supportive role for the brain, and are filled with a fluid known as cerebrospinal fluid. This fluid, which is derived from blood, acts as a cushion and protects the brain—which is an extremely soft substance—from being easily bruised. The current injected by Bartholow would have passed through the meninges and, although attenuated, entered Mary's cerebral cortex. By so doing Bartholow made another important discovery, although it was more or less simply a replication of the earlier work in animals. When the electrodes were placed over the left side of the brain, the electrical stimulation produced muscular contractions on the right side of the body. A jolt of current sent Mary's right arm shooting outward; her fingers splayed out, her right leg kicked forward, and the head deflected to the right. Similar reactions were found when the right side was stimulated, with the evoked movements coming on the left side of the body. (The hole in Mary's skull was big enough to extend over parts of both hemispheres.)

If Bartholow had stopped there, everything probably would have been all right. No harm thus far, and some important information had been gathered. But the trailblazing physician was not yet done.

He inserted an electrode all the way into the brain, into the left posterior part. (A second electrode, which closed the circuit and allowed the current to flow, was placed in one layer of the meninges, the dura mater.)

Here's what happened, in Bartholow's words:

When the circuit was closed, muscular contraction of the right upper and lower extremities ensued, as in the preceding observations. Faint but visible contraction of the left orbicularis palpebrarum [the muscle that closes the eyelid], and dilatation of the pupils, also ensued. Mary complained of a very strong and unpleasant feeling of tingling in both right extremities, especially in the right arm, which she seized with opposite hand and rubbed vigorously. Notwithstanding the very evident pain from which she suffered, she smiled as if much amused.[7]

Bartholow extracted the electrode and inserted it into the right side of Mary's brain. Perhaps frustrated with the slow pace of the investigation, he also elected to quit pussyfooting around: "In order to develop more decided reactions," he wrote in his report, "the strength of the current was increased." Unfortunately, decided reactions were exactly what he got. Again in his own words:

When communication was made with the needles [when the electrode circuit was closed], her countenance exhibited great distress, and she began to cry. Very soon the left hand was extended, as if in the act of taking hold of some object in front of her; the arm presently was agitated with clonic spasms; her eyes became fixed, with pupils widely dilated; lips were blue, and she frothed at the mouth; her breathing became stertorous; she lost consciousness, and was violently convulsed on the left side. The convulsion lasted five minutes, and was succeeded by coma. She returned to consciousness in twenty minutes from the beginning of the attack, and complained of some weakness and vertigo.[8]

Mary never recovered. She kept having seizures, even without electrical stimulation. A few days later, Bartholow visited with her and reported "rhythmical movements of alternate contraction and relaxation of the muscles of the right arm. These soon extended to the shoulder and neck, so that the head moved synchronously with the arm."

Mary's condition rapidly deteriorated. She became incoherent and suffered from seizures and paralysis. She died shortly afterward.

Bartholow concluded his report with a terse sentence: "It has seemed to me most desirable to present the facts as I observed them, without comment." Which was fine, because more than enough comment was soon forthcoming from the rest of the medical community.

* * *

Bartholow's paper describing his experiments on Mary Rafferty was published in April of 1874 in *The American Journal of the Medical Sciences*. It was the lead article of that issue, and thus patently situated to attract attention. It did so.

The reaction was swift and decisive. The medical research community found the results of the experiment interesting but the method of obtaining them was so disturbing that the work was utterly condemned. That very same year the American Medical Association was moved to publish the following resolution: "No member of the medical profession is justified in experimenting upon his patient, except for the purpose and with the hope of saving said patient's life, or the life of a child *in utero*."[9]

The editor of the prestigious *British Medical Journal* published an editorial on the subject in the May 23rd issue of 1874. The editorial begins this way: "An American physician has recently gone a step beyond Dr. Ferrier, and in a direction in which Dr. Ferrier is never likely to follow him. He says that 'nothing has hitherto been done to subject the human brain to a course of experiment in order to determine the nature of its functions'; and presents us with certain researches, after the manner of Hitzig and Ferrier, on the cerebrum of a woman." The editor's indictment is seething but understated. Bartholow's name is never mentioned anywhere in the whole editorial, an obvious omission. Bartholow's very name, like his deed, had become unspeakable. But Bartholow's words are quoted, a full column of them, recounting the most disturbing parts of the experiment.

Bartholow's medical colleagues weren't the only ones who were shocked. The Rafferty experiments provided fodder for the pamphlets of antivivisectionists for decades.

Suddenly the eminent Dr. Bartholow, seemingly impervious and immune to criticism, found himself in serious trouble. He had boldly gone where no man had gone before, and now, in retrospect, he realized he never should have made the journey. But the visions of fame, perhaps, had whitewashed Bartholow's mind, and he had recounted the story of his experimental subject's pathetic fate, in its full and gory detail, "without comment." Only after the howls of protest came did he finally open his eyes and begin to appreciate what he had done.

Assailed from all sides, Bartholow had a critical choice to make. How to respond? There were three options. One, he could simply ignore the criticism and rebukes, severe though they were. Two, he could defend himself vigorously. Three, he could apologize and admit his mistake. To Bartholow's credit he chose number three. More or less.

"Sir," he wrote in a letter to the editor of the *British Medical Journal*, published a week after the stinging editorial, "a case of epithelioma exposing the brain, on which I ventured to make some experiments, has excited unfavourable comment in your widely circulated journal and elsewhere. Under these circumstances, I beg to offer some explanations, which, whilst they do not justify the experiments in question, at least, it appears to me, put the matter in a less offensive shape."

Bartholow made five points by way of explanation. The patient was "hopelessly diseased" and dying was his first point. Secondly, the patient had given her consent. Thirdly, there was reason, wrote Bartholow, to believe that the experiments could be conducted safely; to support this statement he cited the very well known case of Phineas Gage, who had a tamping rod driven through his brain by an explosion but lived for years afterward. Next Bartholow reminded his critics that he had used faradic current, which was not known to damage the brain. (But according to his published report, he had planned on doing an experiment with galvanic current and the only reason it didn't happen was that Mary's health had by this time deteriorated significantly.) Finally, he claimed that an autopsy showed that the cause of death was due to the ulcer; more specifically, it was caused by inflammation and a thrombus (a blood clot).

In closing the letter Bartholow wrote: "Notwithstanding my sanguine expectations, based on the facts above stated, that small insulated needle electrodes could be introduced without injury into the cerebral substance, I now know that I was mistaken. To repeat such experiments with the knowledge we now have that injury will be done by them—although they did not cause the fatal result in my own case—would be in the highest degree criminal. I can only now express my regret that facts which I hoped would further, in some slight degree, the progress of knowledge, were obtained at the expense of some injury to the patient." It was a masterful example of damage control. The editor was appeased, acknowledging the letter with the belief that it would silence Bartholow's critics.

In essence, it did. Bartholow's carefully worded explanation, which admitted guilt though did not go as far as it could have, was accepted. Controversies remained, however. Bartholow was adamant that the experiment did not cause Mary's death, though it seems rather more likely than not that this was wishful thinking on his part. And that he gained the patient's permission before the experiment was vitiated by Mary's abnormally low I.Q.

Still, Bartholow salvaged his career. Although you might think he would have been forced to leave Cincinnati, that doesn't seem to be the case. It certainly didn't happen immediately—Bartholow had been appointed Professor of Medicine at the Medical College of Ohio around this time and kept the job for five years. Only in 1879 did he leave Cincinnati. He moved to Philadelphia in that year after accepting the chair of Materia Medica and Therapeutics at Jefferson Medical College. (Materia medica is an old term for medicine.)

His popularity wasn't hurt either. He continued with his medical practice and his prolific authorship. His well-known book, *Materia Medica and Therapeutics*, eventually went through eleven editions and sold a total of 60,000 copies, making it one of the more impressive bestsellers of that era.[10] In 1881 he also published a widely read monograph titled *Medical Electricity*. (Nowhere in the 262-page first edition does he mention the Rafferty experiments.) On 10 May 1904, he died in his home in Philadelphia. He was eulogized by a positively glowing biography in a journal published by the College of Physicians of Philadelphia.[11]

Long before that, however, in 1876 Ferrier published what was to become one of the most highly regarded works of early neuroscience, a monograph called *The Functions of the Brain*. Ferrier devoted only a few paragraphs to Bartholow's infamous experiment, noting that it confirmed some of the results found from electrically stimulating the brain of animals. Obviously Ferrier felt that Bartholow's contribution of a "slight degree" to the progress of knowledge, to use Bartholow's description, was very slight indeed. Ferrier also made the following comment concerning electrical stimulation of the human brain: "As this procedure is fraught with danger to life, it is not to be commended, or likely to be repeated."

Correct though Ferrier was in many of his assertions, in this he would prove to be mistaken. And quite soon.

# 3
# Exposing the Brain: Early Neurosurgery

Roberts Bartholow's experiment on Mary Rafferty proved several things. One thing it showed is that there's not much that some experimenters won't do to get their name in prestigious journals. Another thing it showed is that if you can get access to the brain, electrical stimulation can do interesting things and the person doesn't feel any pain from small currents. Bartholow took advantage of a cancer that had eroded part of his patient's skull. Although it ended badly, a more restrained procedure could have had a much different outcome. If the cranium weren't in the way, brain stimulation might not be so bad.

The skull consists of 22 bones, eight of them forming the cranium and the rest making up the facial region. The eight cranial bones consist of the frontal bone, the occipital bone in the back, and two pairs of temporal and parietal bones, along with a couple of smaller bones. The temporal bones are found at the sides of the head and the parietal bones make up most of the roof and upper portion of the cranium. Underneath the skull, the surface of the brain—the cerebral cortex—is often divided into areas named after the bone under which it is located: frontal cortex, occipital cortex in the rear, temporal cortex at the side, parietal cortex on top. Each hemisphere of the brain has its own frontal, occipital, temporal, and parietal cortices.

Initially the cranial bones are separate, connected only by tough fibrous tissue. Although tough, the fibrous tissues aren't as hard as bone, and are noticeable as soft spots in a baby's skull called fontanels. This gives an infant's head some much needed flexibility to slip through the birth canal, as well as providing room for the brain to grow. But the bones grow together and fuse quickly, although the skull doesn't completely mature until young adulthood.

A rigid skull helps keep the delicate tissues of the brain safe from the vicissitudes and hard knocks of an active life. It also confines the brain in a container with only so much volume.

Which is a problem if you suddenly find yourself in need of extra room. This can occur in several unfortunate circumstances. Cells normally know when to stop growing and dividing, but occasionally a genetic mutation causes a cell in the brain to continually divide. Neurons don't divide[1], but that's not true for other cells in the brain such as glial cells, which perform various structural and biochemical functions. The result of uncontrolled division is a mass of cells called a tumor whose unchecked growth compresses the rest of the brain. As the pressure rises, part of the brain can even be pushed out through the foramen magnum—the hole where the brainstem meets the spinal cord. In such cases death is inevitable.

Another misfortune can occur after a contusion. A contusion is a bruise, caused by a blow severe enough to damage blood vessels and make them leaky. A bruise on an arm or leg leaves a discoloration that usually clears after a while. This can happen in the brain as well, but a large amount of leaking blood can create a more serious problem in the confined space of the skull. The brain may also swell in response to the injury. All of this can raise the pressure inside the skull to an alarming level.[2]

One way of dealing with this pressure is to give it an outlet. In other words, make a hole. This sounds drastic, and it is, but it's sometimes necessary. And it seems people may have been performing this procedure for a long time.

Archeologists have found skulls with holes that were made a lot more skillfully than with a violent swing of a club or ax. It's called trepanation, meaning to use a trephine—a surgical instrument for removing circular sections, like a drill. But many of these skulls predate modern forms of surgery—they are hundreds or even thousands of years old. Some of the skulls appear to be carefully scraped or cut with some kind of tool, and show signs of healing, which means the patient—or victim—survived the ordeal. These procedures may have been performed for a variety of religious or ritual functions, but some scientists believe at least a portion of them constitute an early form of medical treatment.[3]

Today surgeons use high-speed drills and tiny saws to perforate and cut a section of the skull. This section, called a bone flap, is lifted out of the way, exposing the meninges and the surface of the brain. The process is called a craniotomy. When the procedure is finished the bone flap is turned back over and fastened with sutures or screws. (Sometimes the treatment requires continuous access to the brain and so the bone is removed and not replaced, in which case the procedure is known as a craniectomy.) If only a small opening is required, surgeons may only make a burr hole—they drill through the skull with a tiny drill bit. This is trepanation, or trepanning, although these days most physicians refer to it as burr-hole craniotomy or just craniotomy.

I have a lot of experience with craniotomies and burr holes. Fortunately not with humans—as a postdoctoral researcher I operated on hundreds of anesthetized mice. Opening the skull is rather difficult to do without damaging at least the uppermost layers of delicate brain tissue (unless you have expensive surgical equipment and a lot of skill, which neurosurgeons who operate on people have and I lacked). You also have to dodge a lot of blood vessels; if you hit one, the hole you made will fill with blood. But I needed to perform the surgery so that I could insert slivers of metal electrodes, usually made of tungsten, to record the electrical activity of the brain at specific regions or to deliver electrical stimulation.

Obviously researchers of today can only stimulate the brains of humans in conjunction with medical treatments. Ever since the Mary Rafferty tragedy, neuroscientists have almost always confined their experiments to laboratory animals except for rare opportunities involved in treating human patients for severe brain diseases and trauma.

I say "almost always" because the historical record contains a few exceptions. Little was learned in these experiments—at least the ones that were published—but I want to describe one just to show how desperate and diabolically ingenious brain scientists can get when deprived of sufficient chances to study their object of interest in the species that commands the most attention.

Warning: it's gruesome. Depending on your point of view, it's even more bizarre than the Rafferty episode and nearly as disturbing.

* * *

The experiment was reported in a French journal in 1885 and summarized in the 1 January 1886 issue of *Science*, then as now one of the premier scientific journals in the United States (and the official voice of the AAAS, the American Association for the Advancement of Science). It involved a fresh source of blood along with one dog and one recently decapitated criminal.

The *Science* report fails to indicate if the subject of the experiment volunteered — and it was unlikely to be considered necessary. At the time, experiments on prisoners were more or less condoned by many people in society. Criminals had violated the rules and therefore abdicated their right to be treated with the same respect as the rest of us. Or so went the argument. Hardly an argument that would fly today in the age of custodial rehabilitation, and no doubt it excited unfavorable comment in the past as well. But this experiment had the mitigating circumstance, if it can be called that, of waiting until the subject was no longer connected to his body.

The article begins with an oblique reference to the Rafferty experiment and a reminder of Ferrier's warning: "The problems of brain physiology are so complex, and our means of studying them, especially in the human subject, so insufficient, that it is not to be wondered at if rather out-of-the-way and venturesome experiments are sometimes undertaken by the anxious physiologist; as, witness the actual stimulation of the exposed brain in a patient whose death seemed certain. Such an experiment is not apt to be repeated...."

But a French physician, Jean Laborde, had another out-of-the-way and venturesome idea. Evidently it wasn't a new idea, but the *Science* report states, "the results have been, as a rule, either entirely negative, or brought out only a few rather obvious facts." But in this new and improved version the physicians were careful to maintain "the spark of life, artificially, for a much longer time than was ever before accomplished." Hence the need for the dog and the blood.

A pair of large blood vessels called the carotid arteries carry much of the brain's blood supply, one artery for each hemisphere. You can feel them throbbing in your neck if you put a finger an inch or two on either side of your Adam's apple. To keep the decapitated brain oxygenated, the experimenters pumped blood through these arteries. Into the left carotid the physicians plugged a tube running from the carotid artery of the dog. The other artery received an injection of warmed "defibrinated" blood (a certain protein, fibrin, had been removed so that the blood wouldn't clot).

A person receiving an infusion of dog's blood would have a negative reaction, of course, but for the purposes of delivering oxygen to the brain for a short period of time, the experimenters felt the technique was sufficient. They apparently hedged their bets by using a different system for the other hemisphere.

The experimenters received the head seven minutes after the execution. (The *Science* report gives no details on the victim or his crime.) Another ten minutes passed before they were able to find the carotids due to the "sadly disfigured" condition of the neck. Finally, a few minutes later, blood began to flow: "The result was striking: a bright color returned to the face, which also assumed a natural expression. The effect was most marked on the left side, which received its blood-supply direct from the dog."

I wonder what exactly would constitute a "natural expression" under these circumstances. But I have little doubt that whoever this unfortunate fellow was or whatever his crime might have been, that person—or at least the essence of his being—ceased to exist well before the infusion of blood. The brain had been without blood flow for 20 minutes. It had been shutting down for some time.

Even so, the physicians enjoyed success. They cut a hole in the skull on the left side, near where the frontal and parietal bones meet—under which they believed was the part of the brain responsible for moving facial muscles—and inserted electrodes.

The first time they stimulated they got no result, so they cut another hole a small distance away and tried again. This time they observed muscular contractions on the opposite side, on the right.[4] They even got the lower jaw to move, "causing a strong chattering of teeth." These effects lasted up to 49 minutes after decapitation. After that, stimulation failed to elicit muscular contractions, although the experimenters could still get the muscles to move if they stimulated them directly.

So what was learned from this grisly experiment? Not much. The article in *Science* acknowledges that scientists "have not been very sanguine of results from this method of research." (Sanguine? Was that a play on the word, which can mean bloody as well as optimistic?) But the article cites the exceptional length of the experiment and the means of infusing blood into the decapitated head as important advancements. The idea was to complement the results from animal experiments using whatever source of human tissue became available.

*  *  *

Even as *Science* reported this macabre experiment and re-issued Ferrier's admonition, researchers elsewhere were developing breakthroughs in the art and science of electrical stimulation of the human brain. These advances took place simultaneously with and in relation to the rise of medically successful brain surgery. Or, if the ancients used trepanation at least occasionally for medical treatment as some scientists today believe, perhaps one should say that the late 19th century saw the rediscovery of the art and science of neurosurgery.

One of the most successful pioneers was Victor Horsley, a British physician who was the first person employed in a hospital as a "brain surgeon."[5] Horsley sported a prominent moustache and was something of an eccentric, standoffish person. He studied and worked with some of the most famous British physicians and researchers of the 19th century, including David Ferrier. Like Roberts Bartholow, Horsley made some enemies among his contemporaries for his outspoken manner. But unlike Bartholow, Horsley didn't wade into a great controversy. He seemed to confine himself to helping his patients.

In the 23 April 1887 issue of the *British Medical Journal,* Horsley described ten cases in which he performed brain surgery. He gave the article a rather lengthy title: "Ten Consecutive Cases of Operation upon the Brain and Cranial Cavity to Illustrate the Details and Safety of the Method Employed." He used the word *consecutive* to make sure you wouldn't think he was cherry-picking his results; these were ten consecutive cases of surgery to remove diseased brain tissue arising from trauma or tumors that were causing various neurological impairments such as epilepsy, constant headaches, and paralysis. Only one operation failed.

After acknowledging that he had practiced on animals, Horsley described the important elements of a successful brain operation in humans, including anesthesia and craniotomies. Horsley's preferred method of craniotomy was to use a trephine to test the bone thickness, then take out the required section with a circular power saw. He lifted bone with the aid of sturdy forceps.

Horsley concluded the report by claiming his results "show that the operation of exposing and removing considerable portions of the brain is not to be ranked among the 'dangerous' procedures of surgery." Bold words, even for a man accustomed to expressing his opinion forcefully.

He proved to be right. But removing considerable portions of the brain could have considerably unpleasant consequences. Take too much and you've turned a talking, thinking person into a vegetable. In cases of trauma in which the lesion—the injured tissue—is obvious, you can easily tell what part or parts have to go. In other cases such as epilepsy, it is less clear because a visual inspection often fails to distinguish healthy tissue from that which is causing the seizures—the damage may be microscopic, at the cellular or molecular level. To eliminate the seizures you have to remove the instigator—the focus—but you don't want to go over-board.

This is where electrical stimulation can help. By the late 19th century surgeons began exploring the brains of epileptic patients with their crude electrodes in search of the source of the trouble. Find an area that is particularly excitable and touches off a storm of activity leading to convulsive muscular contractions, and you've found the bad spot. Sometimes physicians were successful and sometimes not so much. During this era electrical stimulation still had something of a Wild West atmosphere. Ferrier, for example, liked to measure the amount of current flowing through his electrode by using his tongue. It is indeed composed of sensitive tissue, but I can't help wondering if Ferrier had trouble getting that metallic taste out of his mouth.

Today's methods are much improved. Treatments have advanced as well. There were only a few effective drugs in Horsley's day, but now most epileptic patients respond to one or more of the considerable number of antiepileptic medications. Happily, in a lot of cases that's the end of the story. But drugs don't help every patient. In such cases, a surface recording—an EEG—or some other noninvasive method can often identify the area, if there is one, at which the seizures begin, and removal of this area generally solves or reduces the problem. Difficult cases such as Brad, described in chapter 1, require physicians to poke electrodes into the brain.

In the process of refining their methods of finding epileptic tissue, researchers began learning a lot about the brain by electrically stimulating it. With the use of small or intermittent doses of anesthesia, surgeons could allow patients to become conscious during brain surgery and describe what they felt and sensed during electrical stimulation.[6] As Bartholow had discovered, the brain contains no nerves to make the tissue itself sensitive.

But of course the brain gives rise to all sensations and perceptions, including painful ones as well as pleasant ones. Patients reported feeling all kinds of odd tingling when current flowed through certain regions of their brain. Harvey Cushing, one of the American pioneers of brain surgery, performed many operations. In 1909 he published a report in *Brain* on two epileptic patients whose seizures were generally accompanied with peculiar sensations in the hand or arm. Cushing went looking for the source of the seizures in a strip of cortex where physicians had previously discovered they could invoke sensations in patients by electrical stimulation. Locating the diseased tissue was problematic, but Cushing did notice that the region of the cortex from which hand and arm sensations could be invoked in these patients was extensive. He wasn't sure if that meant the cortex generally had a lot of area devoted to these body parts or if epilepsy had made the brains of these patients especially sensitive in this area. Subsequently, scientists confirmed that the cortical region responding to sensations in the hands is quite extensive, which seems reasonable—the hands and fingers are highly important in touching and feeling.

Around this time scientists were continuing to study the effects of electrical stimulation on the brains of animals. Although animals couldn't tell you what they were feeling, they were suitable to study motor responses, so you could discover which brain region moved this or that muscle. But because many scientists dismissed studies of rabbits, dogs, and even monkeys as irrelevant to human physiology, British physiologist Charles S. Sherrington decided to use apes. He anesthetized and experimented with chimpanzees along with a few orangutans and gorillas.

Today, researchers at various institutes and universities who use animals must get approval for their experiments from various committees charged with overseeing the use and treatment of animals in the laboratory. It would be exceptionally difficult if not impossible to repeat Sherrington's work with apes today. In the United States the use of animals is heavily regulated, and every once in a while a bill gets brought up in Congress proposing even more regulations.[7] In the United Kingdom, where Sherrington worked, it's even harder. Back when I was working as a researcher I visited a colleague at Cambridge who complained of great obstacles in his research because of the resistance to his use of monkeys.

But in the early 1900s this was feasible. Along with his

colleague Albert Sidney Frankau Grünbaum, Sherrington mapped in great detail the motor region of the cortex—the area of the brain that when stimulated will cause movements in the body. (Early papers from these two researchers list the authors as Sherrington and Grünbaum, but later papers give the authors as Sherrington and Leyton. In 1915 Grünbaum changed his last name, ostensibly because German names weren't popular in England during World War I.) These researchers were exceptionally skilled and careful, and found that the stimulation of motor cortex resulted in complex movements rather than a twitch of a single muscle or two. The researchers also discovered the ability of the cortex to change with experience. The response of the cortex to stimulation could be modified by earlier stimulations—even those that had occurred some distance away. In other words, the cortex sometimes changed its response to stimulation because it had been stimulated before, or because a neighboring region had been stimulated. This, Sherrington realized, could be due to the adaptability of the brain. Later researchers would call this phenomenon plasticity—the ability of the nervous system to change. Learning and memory would be impossible without it.

Cushing, Horsley, Sherrington, and others helped to establish the modern view of the cerebral cortex and mapped its regions. Some parts are involved in motor function, others sensory. The primary motor cortex lies in the frontal lobe, directly in front of a vertical fissure or sulcus known as the central sulcus, which divides the frontal and parietal lobes. This cortex projects mostly to neurons in the spinal cord that innervate and activate the body's muscles. Directly behind the motor cortex, on the other side of the central sulcus, is the somatosensory cortex, which receives information from nerves embedded in the skin that respond to touch and pressure. These two regions—the primary motor and somatosensory areas—are not randomly arranged but are configured in the form of a map of the body. Neurons corresponding to the forearm are next to ones corresponding to the upper arm, which are next the shoulder neurons, then the chest, abdomen, and so on. Facial and hand areas cover a disproportionate amount of territory in these cortical maps because of their importance in human facial expression and dexterity.

Further research identified primary areas receiving information from the other senses: vision (at the back of the brain, in the occipital lobe), hearing (temporal lobe), taste (parietal lobe), and smell (temporal lobe). This information comes from sensors that transmit their signals to the cortex by way of a group of neurons deep in the brain called the thalamus. (An exception is smell, which has a direction projection from the nose to the cortex.)

But scientists discovered that these primary sensory and motor areas cover only a portion of the cerebral cortex. The rest was no-man's land, or association cortex, as puzzled researchers called it. When electrically stimulated, it didn't seem to do much.

But then came an explosion of knowledge that followed in the wake of a neurosurgeon who founded a neurological institute in Montreal. His name was Wilder Penfield.

4
# Wilder Penfield: Flashbacks

If you know anything at all about the subject of electrical stimulation of the brain, you've probably at least heard the name Wilder Penfield. Of all the many people over the years who have contributed to this field, he's certainly one of the best known. This is due in part to the exciting nature of his work and the diligence and tenacity with which he pursued it, but also because he embarked on a second career writing articles, books, and even a few novels once he began winding down his medical and research activities.

Penfield was born in 1891 in Spokane, Washington. Medicine was certainly in his DNA—his father and one of his grandfathers were both doctors. But his father's medical practice didn't prove to be a big money-winner and his parents separated when he was eight years old. Wilder Penfield moved with his mother to Wisconsin. An outgoing, athletic young man, Penfield excelled in school as well as on the sports field and was extremely popular. In 1913 he earned a degree in literature at Princeton University, which undoubtedly came in handy many years later when he began his second career. But first came medicine. Deciding that the world needed improvement and people needed help—a conclusion many intelligent young people arrived at during the 1910s, an era of bitter conflict—Penfield opted to become a physician, despite his father's lack of great success in the pro-fession.

Winning a Rhodes scholarship, he studied for a while in England. Eventually he earned a medical degree in 1918 at Johns Hopkins in Baltimore, Maryland.

Penfield was lucky to fall under the tutelage of some of the greatest minds of the period, including Sherrington and Cushing. An excellent student, Penfield took in a lot of diverse information from physiology, medicine, and surgery. He realized that an optimal program of brain research required knowledge from a number of different specialties[1] (and it still does today, which is why people often refer to neuroscience as an interdisciplinary field of research). In 1928 he accepted a position at McGill University in Montreal, Canada, dreaming of establishing an institute in which a broad spectrum of researchers could study the brain and its diseases.

In 1934 he got his wish. With the aid of a grant from the Rockefeller Foundation along with funds from other donors, Penfield founded the Montreal Neurological Institute. It was here that he performed the work that would make him famous.

By this time, medical researchers had developed many safe and effective local anesthetics. A local anesthetic numbs the tissue around a specific site, such as the injection site of a hypodermic syringe, as opposed to general anesthesia, which causes a loss of consciousness by acting on the brain. Prior to the 20th century cocaine was often used as a local anesthetic, but in the early 1900s researchers began synthesizing similar but safer substances, such as procaine (marketed as Novocain). Numerous others followed. These substances were ideal if you wanted to perform brain surgery on a conscious patient. All you had to do was apply a local anesthetic to the scalp, do the craniotomy and expose the brain, and there you go. Remember, the brain itself has no pain receptors.

One of Penfield's main interests as a medical professional was to treat patients suffering from debilitating seizures. Physicians had already learned that removal of damaged brain tissue can reduce the frequency and severity of seizures in epileptic patients, but this is a drastic step. It is especially problematic in cases of epilepsy involving the temporal lobe, which for some as yet unclear reason is one of the most common areas to be affected. Removal of parts of the temporal lobe is particularly risky because the brain mechanisms underlying language reside here.[2] A language impairment can be even more devastating than the seizures.

To help find the epileptic focus and to prevent the loss of vital brain functions, Penfield used electrical stimulation. This also served to help in another of his interests—discovering how the brain works.

Penfield and his colleagues performed well over 1,000 operations between 1934 and the early 1960s, when Penfield retired (more or less) from his first career. All of these operations involved stimulation of the patient's cortex. And they pretty much covered it all, according to Penfield, who wrote "it is fair to say that, over the years, every accessible part of the cortical mantle of the two hemispheres has been subjected to stimulation at one time or another."[3] Penfield used a 1.5 mm (0.06 inch) silver ball electrode that delivered pulses lasting only a few thousandths of a second at a rate of up to 100 pulses per second. The amount of current was tiny, in the range of about 50 to 500 microamps (a microamp is a millionth of an ampere). When stimulating a site on the cortex, Penfield began with a low current and, if there was no response, gradually increased the amperage.

Most of the time he got no response. But by laborious application of the electrode, moving it from site to site, Penfield kept looking. He placed little sterile markers on the sites to number them and keep track of where he'd stimulated, and used the "landmarks" of the brain to identify areas—a prominent sulcus or gyrus, or perhaps a major blood vessel (many vessels course around the surface of the brain). But the contours of the brain vary among individuals, which was another reason Penfield wanted to test a patient's brain with stimulation before removing any tissue.

In some cases he was able to elicit a mild seizure which, because he was using low current, indicated a spot of overly excitable tissue that could be causing the patient's epilepsy. Sometimes he worked near the motor cortex and he was able to map this area in detail by observing the patient's muscular responses. He also mapped the somatosensory cortex from the responses his patients gave upon stimulation ("I felt something crawling up my leg" or "something brushed my hand").

Remarkable stuff—eliciting fake sensations just by applying a little current to the brain.

All this data permitted Penfield to draw exquisitely detailed maps of the motor and somatosensory cortex of the human brain. Penfield built upon the work mentioned in the last chapter, though sometimes people have the mistaken impression he invented the concept of sensory or motor maps. This is because he presented the maps in a memorable way, with a homunculus—little man—draped over the cortex, with each body part over the region of cortex that corresponds to it. The homunculus has an odd shape because a disproportionate amount of cortex is devoted to the face and hands—the little man has a big head and huge hands and feet but a tiny trunk.

But the most interesting results came when Penfield stimulated the cortex that didn't have a direct correspondence to the senses or movement—the association cortex, or as Penfield sometimes called it, "uncommitted" or "interpretive" cortex.

\* \* \*

In the early decades of the 20th century, most scientists assumed that the cortex which wasn't directly associated with sensation or movement must have some higher function, perhaps related to thinking or problem-solving or an advanced stage of perception, as in seeing the big picture. A comparison of species brings about a pertinent observation: human beings generally possess a lot more association cortex than other animals, especially in the frontal lobe.

So what happened when Penfield stimulated association cortex of these conscious patients? Most of the time the patient felt nothing, but on occasion something really interesting happened. It could take a variety of forms.

Sometimes stimulation created confusion—the patient seemed temporarily mixed up. Which is reasonable, considering someone just threw the electrical equivalent of a spanner in the works.

At other times, patients might make fumbling movements with their hands, or they would shift their body in some way. In a few cases the patient even tried to get off the operating table.

Stimulation in some cases elicited nonsense speech or non sequiturs, such as a patient who suddenly blurted out something about paint.

These strange behaviors had an automatic or robotic quality about them. In many cases the patients had no memory of doing them, as if the stimulation elicited the behaviors without the patient's awareness.

Penfield stimulated one of the temporal lobes in 520 of the operations he performed.[4] Out of these cases, 40 (18 men and 22 women) reported experiencing vivid flashbacks or hallucinations. This was the aspect of Penfield's findings that captured the most public interest, even though it occurred in only a small per-centage—7.7 percent—of patients who underwent temporal lobe stimulation. (The temporal lobe was the only site from which these experiences could be elicited.)

One woman heard the voice of her young son, as if she was in the kitchen and he was outside, in the yard.[5] She also heard other noises commonly associated with the neighborhood, such as street traffic, the barking of dogs, and the voices of other children. When asked if she thought she was experiencing a memory during the stimulation, she replied, "It seemed more real than that." In other words, she said it seemed more vivid than just reliving an old memory.

Some patients excitedly related what they were experiencing: "Yes, doctor, yes, doctor! Now I hear people laughing—my friends in South Africa...." One 12-year-old boy said, "Oh, gee, gosh, robbers are coming at me with guns." Other patients were quiet: "I had a dream. I had a book under my arm. I was talking to a man."

Lamar Roberts, a colleague of Penfield, put it this way: "It is as though a wire recorder, or a strip of cinematographic film with sound track, had been set in motion within the brain. A previous experience—its sights and sounds and the thoughts—seems to pass through the mind of the patient on the operating table." However, the patient knows he's still on the operating table: "At the same time [as the flashback] he is conscious of the present. He re-experiences some period of his past while still retaining his hold on the present. The recollection of the experiential sequence stops suddenly when the electric current ceases."[6]

Although those of us who weren't on the operating table can't possibly know what it feels like, I imagine it might be at least remotely similar to something I've experienced on occasion. Every once in a while in the morning I find myself in a hazy, half-awake half-asleep state. Suddenly I hear a noise—a voice, or footsteps, or a knocking sound—and I'm instantly awake and alert. And then I realize the noise wasn't real. It must have occurred only in my mind. Perhaps part of my brain was in dream mode, or perhaps my auditory cortex detected some kind of disruption in an otherwise steady rhythm and interpreted it as an external noise.

Music was a common theme in the stimulated experiences of Penfield's patients. In a 1968 paper, Penfield wrote: "A good many patients were caused to hear music. In general, each of them seemed to be present while listening to orchestra or voice or choir. The same piece of music could be recalled by the electrode after a short interval and the subject, if asked to do so, could hum along, accompanying the piece that he was hearing."[7]

Which raises an interesting question. Was the same experience always elicited each time Penfield stimulated the same spot of cortex? The answer: sometimes but not always.

Some degree of variability isn't surprising for these experiments. Even the tiny electrode tip used for stimulation was dozens of times bigger than most neurons. The electrode must have sent current through many cells in a complicated fashion, causing some to fire off an action potential or perhaps a series of them, while other cells may have been inhibited. The action potentials went racing down the axons and triggered the release of neurotransmitters onto other neurons across synaptic junctions. An avalanche of complex activity would have cascaded through the patient's neural circuits. A slight shift in the electrode's position, or a change in current, or even just fluctuations in the complicated way charges move through biological tissue could result in different responses.[8] Subsequent stimulations would be unlikely to reproduce exactly the same electrical activity as the first.

Patients also experienced what Penfield referred to as interpretive illusions. Unlike hallucinations, which replay scenes from a different place or time, interpretive illusions are distortions of the present. For example, sounds may be amplified, or they may be reduced, seeming to come from a great distance. Objects may appear at greater or lesser distances than they really are, and accordingly might be smaller or larger, or perhaps blurred, flickering, or in an unusual state of motion. There could be illusions of eerie familiarity—déjà vu. Sometimes emotions such as fear or disgust could arise.

In the patients undergoing electrical stimulation, clearly the elementary sensations elicited from sensory areas and the interpretive illusions elicited from association cortex were products of perception or altered perception. But were the "flashbacks" actual memories?

In some cases patients experienced something familiar—something they had experienced many times in their lives, like the woman who heard her son playing outside. In other cases, though, the scene or snippet that popped into their minds was something that probably never occurred in real life, such as the young boy's vision of robbers coming to get him. Some patients referred to their stimulated experiences as dreams and other patients called them flashbacks. But they all agreed that whatever they actually were, they were generally more vivid than any sort of memory that they might recall when they were thinking about the past.

Penfield believed all of the episodes resulted from the activation of some sort of memory trace, although not necessarily an actual event—it could have been from a dream the patient once experienced or from something in a magazine, book, or movie. In the cases where the events of the flashback could be checked for historical accuracy, they proved to be true memories.

There is another important question to ask about these findings: do they apply to people without epilepsy? Many of the patients who experienced flashbacks or dreams during electrical stimulation also did so during their seizures—temporal lobe seizures commonly invoke odd sensations or mental phenomena. In some of Penfield's patients, the stimulation-induced flashback and the seizure-induced flashback were identical; in other words, the stimulation seemed to evoke the same episode as the seizure (and the stimulation sometimes did generate seizures in the patient). But in other cases it was different. And some patients who had experienced flashbacks during their seizures didn't experience any from the electrical stimulation. So there wasn't a precise correspondence between seizure experiences and electrical stimulation experiences in all patients.

Would mental phenomena similar to these patients happen in non-epileptic brains? Penfield certainly thought so. Although he acknowledged that epilepsy makes brain tissue easier to excite, or in other words decreases the current needed for activation, Penfield argued that epilepsy didn't transform the basic processes of brain function: "This increase in stimulability (decrease in threshold) does not mean that the epileptic process is responsible for the nature of the response. It is clear enough that we are activating a normal mechanism of the brain...."[9]

But nobody really knows for sure.

* * *

In the last quote from Penfield cited above, I cut him off. It's not that what he wrote next was irrelevant, it's just that I wanted to use it to begin a new section. Penfield was arguing that his research was applicable to everyone, not just epileptic patients, and then he proceeded to discuss how it helps us understand how the brain works. Let me finish the quote: "...and after this revision of our material [he's referring to the article he was writing, in which he and a colleague summarized data from their previous work], we should try to understand how the mechanism is employed in normal living." That's what I want to do in this section.

What Penfield did was inject current into the brain of his patients and set into motion some of the machinery of their cognitive processes. In many cases the results of those processes were either nonsense or failed to reach conscious awareness, and in other cases the stimulation sparked a mere chirp or flash of light or muscular twitch, depending on where the current was injected. But with temporal lobe stimulation, Penfield occasionally elicited some kind of memory or flashback.

Why those memories and not others? Most of these memories didn't seem highly significant, and a lot of them were trivial. Penfield believed that his electrode called up random incidents from the collection of memories that the patient had stored. Since there was nothing special about most of these memories, this gives rise to the speculation that all or nearly all of a person's experiences have been retained. If this is true, every little incident is inscribed somewhere in your brain.

Science fiction writers have occasionally made use of this hypothesis. Once neuroscientists gain a better understanding of the brain, or so the stories go, maybe we can pull memories out of brains like a video from the internet. Even today, hypnosis is often used in the attempt to dredge up memories that aren't consciously accessible. I have my doubts about hypnosis, but in the future, electrodes might offer a more fruitful alternative.

But the premise of unabridged memory strikes me as fanciful and unlikely. Consider this: a single day consists of 86,400 seconds. It's true that events don't happen every second and we're sleeping about a third of the time, but if you're going to record everything, I guess you'd need at least 40,000 seconds of storage space. Multiple that by 365 days and then some number of years, and the result is a large number even for a young person. It's possible that the brain can hold this much data, and perhaps it does. The brain probably evolved to store all information pertinent to survival, and because it's not always immediately obvious which information is pertinent, the brain may store it all, at least for a while, so that it can be reviewed later as necessary. But I suspect there's a dumping mechanism too, otherwise there would be a lot of clutter.

We can also ask how memories are stored. That's a big topic in neuroscience but not the main thrust of this book, so I'll leave it alone except to say that most neuroscientists believe that changes in synapses underlie many forms of learning and memory. The idea is that a synapse becomes stronger or weaker, which affects the route and flow of electrical activity in the brain. The memory is stored in the synaptic weights, or strengths. According to this hypothesis, events leave a trace—and ideas become associated in the brain—because changes in synaptic strengths alter the way information is processed.

Another big question: *where* in the brain is memory stored? If the synapse theory is right, where are these synapses located?

You might be tempted to conclude from Penfield's work that memory is located in the cortex, specifically the temporal cortex. Stimulate the temporal cortex and sometimes patients have flashbacks.

But remember that the activity spreads beyond the area of stimulation. The action potentials and the activity these signals created farther downstream may have gone anywhere in the brain. And Penfield discovered that the cortical site of stimulation wasn't essential for the memory. If that section of cortex was later excised as part of the treatment for epilepsy, the memory it had invoked remained. Penfield assumed this meant that the cortex wasn't the seat of memory.

So what does cortex do? Penfield had some ideas. Consider, for example, the experiences that were commonly elicited during electrical stimulation, and then consider those that weren't. According to Penfield, "The times that are summoned most frequently are briefly these: The times of watching or hearing the action and speech of others, and times of hearing music. Certain sorts of experience seem to be absent. For example, the times of making up one's mind to do this or that do not appear in the record. Times of carrying out skilled acts, times of speaking or saying this or that, or of writing messages and adding figures— these things are not recorded. Times of eating and tasting food, times of sexual excitement or experience—these things have been absent as well as periods of painful suffering or weeping."[10]

Penfield speculated that the association cortex was for inter- pretation (hence the name he sometimes gave it), elaboration, and comparison of information. Its function isn't simply to indicate the presence or absence of a noise or object, or just to store memories of such sensations or to convey their emotional content, but rather to compare what is present with previous experiences of these things. The simpler functions, such as anger or happiness, must be found elsewhere. (That's the subject of the next chapter.)

As far as memories go, researchers have since identified a structure that seems to be vital for laying down new memories, and Penfield played a role in this discovery. One of the earliest indications that this structure, known as the hippocampus, was crucial for memory came from operations in which it was removed as part of the treatment for epilepsy. The hippocampus looks kind of like a seahorse, hence its name (which means seahorse in Greek), and is located in the temporal lobe. There is a hippocampus in each hemisphere, so people have two of these structures. In 1953 a physician named William B. Scoville removed both in some of his patients, and as a result, patients experienced severe anterograde amnesia—their brains didn't form any memories of new experiences. Their memories became frozen at the moment of their operation, without any updates of what they experienced since then.

One patient, known by the initials H.M. (Henry Molaison), was initially studied by Penfield and a colleague, Brenda Milner, because Penfield and his staff had a lot of experience evaluating such patients. Subsequently H.M. and his personal tragedy became well known. After the operation he lived every moment in the present, with no memories except those that occurred before the operation. If you were to introduce yourself to H.M. and slip away for a moment, when you returned he wouldn't remember meeting you. As he aged he became horrified when he looked in the mirror and saw an old man—to the best of his recollection he was still 27 years old, the age at which he'd had the operation. In 2008 H.M. died.

Although the hippocampus clearly plays an important role in the development of memories, many neuroscientists believe that the cortex is important in the storage and retrieval of conscious memories (as opposed to say, learned skills, which seem to be maintained elsewhere). This is true despite Penfield's finding that excision of the site of stimulation didn't delete the memory—there may be many neurons and synapses involved, which may spread the memory over vast swathes of cortical (and perhaps subcortical) territory.

What about consciousness? Penfield's electrode elicited conscious, vivid experiences. Certainly you would expect the cortex, with its complex circuits and its expanded territory in the "higher" animals and humans, to be involved in what many people regard as the highest function, conscious awareness. Surprisingly enough, Penfield downgraded the role of cortex because he thought the cortex wasn't where everything came together. According to Penfield, the brainstem played a central role because of its interconnections with the cortex and its role in critical functions of the body. That's not to say that Penfield believed consciousness resides in any one place, but he did believe that evolutionarily older structures—the stuff beneath the cortex—were critical components of the process.

While Penfield's ideas on consciousness are not widely discussed these days, nobody can deny that electrical stimulation can invoke a memory, flashback, hallucination—a thought of some kind. And the patients often described it as more real than an ordinary memory. What is it about electrical activity in the brain that creates consciousness? Are thoughts directly related to electrical activity in a sort of one-to-one correspondence? Does a specific pattern of activity produce a specific thought, and another pattern produce a different one?

I'm not sure how that would work. New thoughts would have to be generated by new patterns. But people remember things too, and we remember that we had this or that particular thought before. So every time you have that thought, does it create a slightly different pattern of activity? Add one neuron, subtract another one, or reduce the role of yet another?

Why do some stimulations lead to conscious phenomena, while other stimulations don't produce anything? What is the mechanism by which the stimulated activity sometimes intrudes into awareness?

Nobody knows. These are questions that will arise throughout this book.

Penfield summarized his early work in 1950 with the book *The Cerebral Cortex in Man*, and wrote several other fascinating books during his second career. He died in 1976, after a long, productive lifetime. He received many awards and honors from his adopted country of Canada.

In case you're wondering, Penfield reported excellent to satisfactory medical outcomes in many of his patients: "Approximately half of the patients would consider themselves cured by operation, some with continuing medication and some free of all medication. Still others were improved and not cured. There were bitter disappointments, but in general this is a most rewarding therapeutic field, though difficult."[11]

But enough of this lying around on the operating table. What if you could implant electrodes in the brain of an awake, unrestrained animal or person? Could you change behavior? Could you invoke certain behaviors? With these questions we move closer to the promise, and perhaps peril, of brain stimulation.

# 5
# Walter Hess and James Olds: Rage and Reward

The development of local anesthesia and advances in instrumentation as well as the increasing skills of surgeons such as
Horsley, Cushing, and Penfield in the early 20th century gave
researchers access to the brain, and the mind, of a person. But the
electrical stimulation of the surface of the brain—the cortex—
produced a limited range of responses in patients. The stimulation
could mimic elementary sensations such as a noise or a flash of
light, it could distort perception, it could set in motion some sort
of automatic program such as hand movements, and in certain
cases it could elicit a flashback or memory. As Penfield himself
noted, lots of behaviors are missing from this list, including
feelings of hate and anger or love and pleasure, which were rarely
if ever invoked.

But the brain is responsible for these raw emotions just as
much as it is for our sense of sight or hearing and our ability to
think, use language, and plan for the future. To study behavior at
a more fundamental level, researchers turned to animals.

Researchers had been using electricity to study the nervous
system of animals even before Luigi Galvani wrote about animal
electricity in the 18th century, but the subject kicked into high gear
with the work of German physiologists Eduard Hitzig and Gustav
Fritsch in 1870. As mentioned in chapter 2, these researchers
elicited movements in a dog by electrically stimulating its cerebral
cortex. The animal was not anesthetized—this was a crude and,
most people today would say, prohibitively cruel experiment.
Subsequent experiments in the decades that followed generally
decreased the suffering of the animals, but retained the simplicity
of the original: stick an electrode somewhere in the brain, turn on
the current, and watch what happens.

Problems with this approach include the inability of an animal
to tell you what this stimulation "feels" like. But that's not an

issue if it's behavior you're trying to stimulate rather than some sort of abstract thought. A problem does arise, however, if the behavior you're trying to elicit is governed by a structure beneath the surface of the brain.

Suppose you've done a craniotomy of an anesthetized subject and you see the surface of the brain. Great, you're all set—if the cortex is what you're looking for. If not, you're in trouble because you can't see below the surface.

All isn't lost, though. Anatomists have studied the structure of the brain and have produced detailed illustrations of its interior, so you've got a general idea where to look. The cortex consists of six layers of cells of different varieties: some are support cells called glia, some are small neurons, and others are large neurons that have branches known as dendrites sticking out of them, making them look like trees (see note 6 of chapter 1). Beneath the cortex runs the wiring—the axons of neurons crossing long distances (relatively speaking), carrying action potentials of neurons in one part of the brain to another. Many of these axons are sheathed in a fatty substance known as myelin which speeds the conduction of the action potential and makes it more efficient. Myelin imparts a whiteness to the axons, so the wiring of the brain is generally white. Cells are mostly transparent but sections of tissue with a high cellular content can take on a grayish hue, so cells are sometimes called "gray matter," as opposed to the "white matter" of myelinated axons.[1]

The brain has a lot of white matter because neurons do a lot of communicating. But buried amidst the white matter in the interior of the brain you can find groups of neurons that form little enclaves. The generic term for a group of these cells is nucleus, such as the caudate nucleus and the subthalamic nucleus. These nuclei perform specific functions or calculations, often of a sensory or motor nature, or they may be involved in basic phys- iological processes such as respiration or sleep. Many of these nuclei have been subdivided based on their function or structure; often the neurons in a nucleus will project their axons to specific parts of the brain, and sometimes a subset of these cells will project to one region while another subset communicates with a different region. It's all part of a grand complicated scheme neuroscientists have just begun to decipher.

If you're aiming for a subcortical nucleus, you usually know the general vicinity of the target, but aiming a needle-thin electrode into the brain is rather like tossing a dart at a board in a totally dark room. You know the target is on the wall but you're not sure where. So you throw the dart and hear it thud into something. And then you wonder—what have I hit?

With an average volume of about 1,500 cubic centimeters (91 cubic inches) and a target of perhaps a thousandth of this volume or even less, you might not come close. The current you send through the electrode may not travel anywhere near the target. Worse, it might activate something entirely different than you intended.

To solve this problem, researchers began to make three-dimensional maps of the brains of a variety of species. In general this is known as a stereotaxic procedure.[2] Locating a point in a solid body requires three coordinates. If the object is roughly spherical, one way to locate a point in the interior is to specify a spot on the surface—that takes two coordinates (think of latitude and longitude on the Earth's surface)—then you indicate the depth at which you can reach the desired point. (The penetration should be perpendicular to the surface.)

The skull conveniently comes with its own landmarks, called sutures. These, you may remember, are where the individual bones of the skull meet and join. They act as latitudes and longitudes. To locate a particular subcortical region, you find the correct point on the surface and drill a burr hole or do a craniotomy, and if you know the depth to insert your electrode then you're good to go. If you're dealing with the surface of the brain rather than the skull you have your choice of landmarks, such as a specific sulcus or gyrus.

This procedure requires a precision instrument because you don't have a lot of room for error. You have to hold the head in place, and you need a frame and some mechanical attachments or arms that can be adjusted with micrometer accuracy. One of the earliest and best instruments came from the excellent surgeon and innovative researcher Victor Horsley, along with his colleague Robert H. Clarke. In 1908, these two men published the specification for what became known as the Horsley-Clarke stereotaxic instrument. You've probably never heard of this device even if you're an avid reader of books about the brain, but without such tools neuroscientists would know almost nothing about the vast region of the brain beneath the cortex.

Researchers got busy making maps and plotting coordinates of animal brains. They inserted electrodes, marked the area of the brain at the tip (by injecting a dye or burning a small hole in the tissue by cranking up the current), and then sliced and stained the brain to pinpoint where the electrode had gone and which subcortical structure it had hit. With these stereotaxic maps and guides, researchers had a good chance of hitting what they were aiming for.[3]

A Swiss physiologist by the name of Walter R. Hess zeroed in on a tiny but critical part of the brain called the hypothalamus. Then he developed an ingenious technique of stimulating it in unrestrained animals and observing their behavior. The result was pretty much everything Penfield didn't see.

* * *

Born in East Switzerland in 1881, Hess's father, a physics instructor, let him play with scientific equipment when Hess was a boy. But Hess didn't begin his career as a physiologist; like many researchers who study the brain, Hess started out in a different field. He obtained his medical degree from the University of Zurich in 1906 and became an eye doctor. As an ophthalmologist he learned to wield precision instruments, a skill he would use later in the laboratory. After finding success as a doctor and earning a bit of money, Hess chose to abandon his practice and go back into training, this time as an assistant to a physiologist. It was a financial burden but the pull of science was too great.

He chose Bonn, Germany to study physiology. This was in the

early 1910s, which wasn't a propitious time to be in Germany, but Hess left before the turmoil of World War I. In 1917 he became the Director of the Physiological Institute in Zurich. He was interested in basic physiological functions such as blood flow and respiration, including the mechanisms that regulate them. And that got him interested in the brain. Ultimately Hess decided to see what sort of behaviors he could invoke by stimulating some of the deep structures in the brain that appear to govern fundamental processes in the body.

But first he had to develop a method. Like most scientific pioneers, Hess spent years perfecting his technique (which, thanks to his efforts, other researchers can now perform with ease).

Hess used cats in his experiments. A cat is a good sized animal for this kind of experiment. A mouse is too small to cope with the weighty electrical apparatus required for these experiments. Subcortical structures in a mouse's tiny brain are also hard to hit, and you can't fit too many electrodes into such a small space. A monkey is big enough, but it's hard to handle and, not being domesticated, will raise objections on general principles to human company. Although cats have been used in a lot of neuroscientific research on vision—they have excellent vision and their visual system resembles a human's—today rats have replaced cats for many of the kinds of experiments Hess conducted.[4]

With cranial sutures as landmarks, Hess and his assistants carefully measured the proper depth to insert their electrodes. The electrodes were made of steel with a diameter of about 0.25 mm (0.01 inches)—they had to be extremely thin so that they would cause as little damage as possible when they pierced through the cortex and the rest of the brain. Steel conducts electricity but the whole electrode wasn't a conductor because the electrodes were electrically insulated except for the tip; only the tip delivered current. Electrodes were inserted in pairs. With paired electrodes you can limit the spread of current by using one electrode as the positive terminal and the other as negative. In this arrangement, the current flows between the electrodes, exciting only the cells in the immediate vicinity. (If you use just one electrode for stimulation, the current will spread out and flow back through your electrical equipment via a ground electrode you've attached somewhere, such as the surface of the brain. You must have a complete circuit or current won't flow.) Hess made sure he was stimulating the right area by sacrificing the animal after the experiment and sectioning the brain to locate the electrode track.

Hess and his team became astonishingly proficient in this operation. According to his student and colleague Konrad Akert, the operation took about 20 minutes. Once the anesthesia wore off—after about 30 minutes—they were ready to start the experiment.

In 1932 Hess wrote all of this up in a book. But he wrote it in German. In fact, most of Hess's early work was published in this language. This choice of language limited his audience, even though many scientists of this era were required to learn several languages—and physiologists almost always learned to read German since so many of the great physiologists of the 19th and early 20th centuries were Germans.[5] But the 1920s and 30s were a time of great nationalistic animosity, and many British, French, and American researchers refused to read anything written in German except when they were forced to. (So much for scientific objectivity.)

During the stimulation experiments the cats were awake and unrestrained, although electrical wires had to run from the electrodes to the current generators. Hess built a rig where the wires were secure but had some slack to allow the cat to move around.

The first few experiments produced inconsistent results. Hess was at the cutting edge of brain science in the 1920s when he began his experiments, and such things rarely work the first time. He called his early work "a thorough disappointment."[6]

But his skill and precision improved. Hess discovered that you could elicit all kinds of behavior by stimulating various parts of a cat's hypothalamus and related structures deep in the brain. The hypothalamus sits at the base of the brain, underneath a somewhat large nucleus called the thalamus. Neurons of the hypothalamus control a great deal of hormonal activity, and the hypothalamus plays an important role in many kinds of fundamental behaviors such as eating, aggression, and so on.

When the experimenters stimulated a certain part of the hypothalamus the cat behaved as if it had just seen a dog. Even a quiet, well-behaved cat would suddenly turn unruly. It adopted a defensive posture, its hair stood up, it hissed, and it retracted its ears. Anything approaching the animal would meet with a fierce attack. But once the current stopped, the aggressive stance disappeared at once. The cat did, however, appear to be in a bad mood for a little while longer—its tolerance for unpleasant stimuli was much lower.

One wonders what the cat was experiencing when the current was turned on. The cat's behavior was accompanied by physiological changes such as the release of epinephrine (also known as adrenaline) seen in the fight-or-flight response, which is switched on during times of great stress. Yet there was no stressor. If you could have asked the cat what it was feeling, what would it have said?

Some of Hess's contemporaries questioned whether the cat was feeling anything at all. In other words, the stimulation evoked the behavior without any sort of emotion or awareness, somewhat like the automatic behavior Penfield occasionally elicited from his patients. In this hypothesis, the cat would have said, "Rage? What rage? What are you talking about? I was just sitting here minding my own business." One researcher repeated Hess's experiment and claimed to observe cats that groomed themselves or ate at the same time they showed this "rage." The behavior elicited under hypothalamic stimulation was therefore called sham rage.

While Hess and his colleagues admitted that electrical stimulation may not mimic normal function, they argued that the rage was in some sense real. Certainly the behavior was real. Akert wrote, "Skeptics may convince themselves of this unambiguous behavior by being exposed to the ferocious and well-directed attack with teeth and claws."[7] One suspects Akert was writing from experience. Today the term "sham rage" is often used to describe this behavior though only because it was elicited electrically rather than by a real dog or some other provocation, not because of what the cat is feeling or not.

Although nobody knows what the cat experienced, it seems they didn't like it. Experiments conducted by other researchers in the 1950s gave cats an option of avoiding the hypothalamic stimulation by, say, jumping over a wall. When the cats learned how to avoid the stimulation, they did so.

In addition to rage, stimulation can also elicit eating and drinking behaviors, elimination, stress responses, sleep or resting states, and heat-dissipation responses such as sweating, panting, and licking. Most of these sites are hypothalamic or nearby. These sites are generally quite close, which is probably why some researchers get a mixture of behaviors during stimulation—their electrodes are almost but not precisely in the right spot.

Hess received the Nobel Prize in physiology/medicine in 1949 for his work. It proved to be one of more controversial Nobel selections, though not because of the quality of Hess's work—it was first rate—but rather because the experiments made some people uneasy. What if you could evoke this sort of thing in people?

Another reason for the controversy was due to the person who shared the physiology/medicine prize that year: Egas Moniz. Nobel committees may select up to three people to share a single prize; sometimes they choose colleagues who have worked closely together, but on other occasions the selection is based on a kind of theme, such as the biochemistry of a certain molecule or the physiology of a certain system. In 1949 the theme appeared to be Manipulation of Behavior by Questionable Means. Okay, maybe not, but Egas Moniz was an interesting choice to say the least. Moniz was a Portuguese neurologist who studied and promoted the operation known as frontal lobotomy, performed on patients with various psychiatric problems. This operation, which severs the connections of the frontal lobe to the rest of the brain, tends to produce docility in otherwise violent patients. It also, according to critics, tends to rob the person of a great deal of what we would call personality as well as cognitive processes such as the ability to plan for the future.

Hess turned his attention later in his career to big questions such as the nature of consciousness and how it arises from brain activity. Like most stubbornly empirical biologists who tackle this issue, he found it something of a quagmire. Even though scientists can record electrical activity of the brain and modify it with stimulation, how the brain's activity gives rise to thoughts is entirely mysterious. Hess ended a 1964 article in *Science* with this sentence: "While neuronal patterns determine the content of consciousness, they fail to provide clues concerning the transformation of such patterns into subjective experience."

Like Penfield, whom Hess admired, Hess tirelessly devoted himself to his research or communication thereof. Also like Penfield, he enjoyed a long and productive life. Hess died in 1973.

* * *

At the other end of the spectrum from rage is a feeling of serenity or pleasure. Electrical stimulation of these feelings began to garner a lot of publicity with the work of a researcher by the name of James Olds.

Recall how tough it is to aim an electrode and hit something in the brain hidden beneath the cortex. One day in 1954 Olds missed. He was doing experiments on rats and was aiming for a group of neurons deep in the brain called the reticular system, which had earlier been shown to be involved in an animal's state of alertness or sleepiness. But he was just a little off.

In this series of experiments, Olds and his coworker Peter Milner implanted electrodes and cemented them into place with a plug that could be connected and disconnected to the stimulator. During the days or weeks that the experiments took place, the rat could be moved around or given periods of rest away from the electrical equipment.

In one trial, Olds placed each rat in a large box and labeled the four corners A, B, C, and D. The animal was put into the box and it roamed around, exploring, as rats usually do when they encounter a new place. Whenever the animal wandered into corner A, the experimenters turned on the electrode current to see if it would have any effect on the animal's behavior. While performing this experiment on the animal with the misplaced electrode, the experimenters discovered this animal continually returned to corner A during the first day of the experiment.[8] It lingered in A even longer at the beginning of the next day.

Penfield failed to find anywhere in the cortex where electrical stimulation evoked emotions such as pleasure. Many scientists in the 1950s believed that such emotions probably involved most or all of the circuits of the brain. If this is true, eliciting pleasure from a localized electrode would not be possible.

Olds believed that the stimulation was likely producing a state of curiosity in the rat. But then Olds got creative with the experimental paradigm. "On the second day," he explained, "after the animal had acquired the habit of returning to corner A to be stimulated, we began trying to draw it away to corner B, giving it an electric shock whenever it took a step in that direction. Within a matter of five minutes the animal was in corner B. After this, the animal could be directed to almost any spot in the box at the will of the experimenter."[9]

What was going through the rat's mind? Maybe it was something like, "Oh yeah, that feels as good as eating a heaping bowl of cheese." Or maybe it was something less specific: "It seems nice and safe here, think I'll stick around." Or maybe he couldn't have explained it: "I don't know why, I just like it better in this corner."

Or perhaps this question is meaningless. A rat isn't a person and doesn't have the same cognitive faculties. Maybe nothing went through its mind because it doesn't have a mind.

It's not my intention in this book to enter into a lengthy discussion about the degree to which consciousness lights up the subjective lives of nonhuman creatures. It's my opinion that most mammals have at least the rudiments of awareness if not something close to our own inner life, but I'm not going to spent time justifying this position. Asking what was going through these animals' minds is, to me, a fair question, even though it's one we can't answer.

Olds found that a rat will even prefer stimulation over food. When a rat that hadn't eaten in 24 hours was presented with food but received brain stimulation only when he was some distance away, he opted for stimulation.

To measure the strength of the stimulation's effects, Olds and his colleagues decided to let the rat "tell" them how much it wanted the stimulation. Doing this required the use of tools that psychologist B.F. Skinner had developed. Skinner was keenly interested in quantifying behavior (mostly because he didn't put any stock in mental constructs or questions such as the one I posed about the rat's mind—he believed behavior was the only thing in psychology that could be studied). In his studies of reward and reinforcement, Skinner built cages in which some kind of action, such as the pressing of a lever, was required to receive a small reward such as a morsel of food or sip of water. With this technique he quantified how hard an animal would work for certain rewards. Olds adopted this approach, but instead of combining a lever press with food or water, he configured the lever to complete a circuit and deliver a small current into the rat's brain. This allowed the animal to stimulate itself—self-stimulation. While the rat was in the box (sometimes called a Skinner box), the experimenter could judge how much the animal would work for the stimulation by how frequently it pressed the lever.

The electric reward was a one-second pulse of alternating current at 60 cycles, just like what you get from an electrical outlet in the United States, only stepped down to a few volts. Press the bar, get a second-long jolt. Some animals pressed for 24 straight hours, as often as 5,000 times an hour.

It's reminiscent of an addict. Think of drug abusers or alcoholics who know, at least intellectually, they're destroying themselves, but can't quit.

Olds and other researchers got busy and mapped the brain to find the areas in which the rat would self-stimulate. There turned out to be a cluster of sites, but one of the most important was a section of the hypothalamus. Rats would also avidly self-stimulate regions located at various points within the brain where neurons project to the hypothalamus or receive information from it. Important regions include parts of the brainstem and the septal area, which consists of a group of small nuclei buried deep in the brain and involved in the regulation of emotion. Particularly potent is a group of axons known as the medial forebrain bundle that carries information between these sites. By conducting exhaustive surveys of the parts of the brain involved in processing emotions, Olds claimed roughly a third was rewarding, a few percent was noxious, and the rest was neutral.

The irresistible question to ask is whether these results apply to humans. At first Olds thought so, or at least hoped so. He ended a 1956 article by noting some experiments performed in monkeys in other labs had replicated the results he'd gotten in rats, which meant "that our general conclusions can very likely be generalized eventually to human beings—with modifications, of course."[10]

But controversies soon arose. Does the electrical stimulation excite the same circuits as a "natural" reward such as sex or chocolate? Many scientists began criticizing the term "pleasure centers." How pleasurable was the stimulation? Yes, the rat would work for it, which meant it was in some sense reinforcing, but was it genuinely pleasurable? People sometimes do things compulsively without deriving much enjoyment from it—many drug addicts, for instance, don't seem to feel pleasure in their continued drug use as much as relief from agonizing withdrawal symptoms if they try to quit.

What's more, sometimes a rat will work just as hard to turn off the stimulation that it had worked to turn on! You think he's loving it, but if you reconfigure the equipment to give him a large dose of it, all of a sudden he hates it. But maybe that's to be expected. A pleasurable sensation is often limited and constrained; I love chocolate but I know better than to eat it nonstop.

The word "centers" also raised some eyebrows. The pathway by which these regions are connected seems to be the most vital element. It's more a circuit than a collection of centers.

Olds' career was cut tragically short when he died in a swimming accident in 1976. But long before he died he had grown cautious about his discoveries and their applicability to humans. He didn't believe electrical stimulation of the "pleasure centers" had much promise or future as part of any sort of therapy or experiential thrill-seeking in humans. In a 1973 book, science journalist Maya Pines quoted Olds as saying: "We know only where a mixed bag of emotional fibers happens to be passing, where we can induce a rather sickly pathological condition."[11]

Caution was not a characteristic of some of the researchers we'll meet next.

# 6
# José Delgado and Robert Heath: Shaping Behavior

The front page of the *New York Times* for Monday, 17 May 1965, was filled with the usual reports: an attack on a U.S. air base in South Vietnam resulted in the death of 21 Americans and damage to 40 planes; the defection of a Polish diplomat; a "teach-in" on President Johnson's Vietnam policy; distribution of children across schools in New York City to achieve racial balance. There was also something unusual: a picture of a "matador" waving a cape at a bull. But the bull had electrodes in his brain and the "matador" was a researcher at Yale University's School of Medicine. His name was José Manuel Rodriguez Delgado.

If Penfield is one of the most famous names in the history of electrical stimulation of the brain, Delgado is perhaps the most *in*famous. Any self-respecting website devoted to revealing the "truth" about the efforts of governments or evil corporations or mad scientists to control our minds will mention Delgado's name somewhere in all those conspiracies. He pioneered many of the modern tools and techniques of brain stimulation. And he was not, by any means, a shy man.

And this was the go-go era of science. Concomitant with the angst of the cold war, civil unrest, cultural upheavals, and 'Nam, scientific research was bubbling with energy and a little angst of its own. After the huge impact of physics in World War II—the atomic bomb is the most obvious example, but there were other major contributions—government officials in the United States decided science was too important to leave to chance. Prior to WWII, researchers had to scrape up funds from their universities, and often used their own money, to equip their laboratories. Afterward, the government picked up the tab, handing out grant money for projects deemed worthy by government agencies and panels of scientists. And then in 1957 the Soviets launched

Sputnik, the first artificial satellite, which displayed their early lead in the space race. The U.S. government poured bucketfuls of money into science and engineering in the frantic effort to catch up. Throughout the 1950s, 60s, and early 70s most research projects got funded no matter what they were. Scientists were hard-pressed to come up with enough ideas to use all the money that was available.[1]

The tons of money flooding into science meant that a lot of interesting if not wholly odd research programs received support. Some of the more intriguing studies explored the extent to which electrical stimulation of the brain can adjust behavior and mental states. The scientists were ostensibly aiming for practical applications such as new therapies to treat mental illness as well as a better understanding of how the brain works. Funding agencies might have had other ideas—this was an age in America history of great unease and fear of communist infiltration, and perhaps people were worried that electrical stimulation of the brain might prove a useful tool for communist brainwashing.

So into this cash-fueled cauldron of angst and science and fear walked Delgado. He fit in nicely: brash and energetic himself, he was eager to learn what his predecessors such as Walter Hess had discovered, and then take the baton and run with it as far as he could. He got pretty far.

Delgado was born in Ronda, Spain in 1915. He received his medical degree at Madrid University. In 1940 he took a professorship in physiology at the Medical School and the Cajal Institute in Madrid (named after the famous Spanish anatomist Santiago Ramon y Cajal). Ten years later he came to the United States and accepted a position in the Physiology Department at Yale University. It was here that he embarked on his quest to find the neural basis of behavior.

When I was a postdoctoral researcher I got deeply interested in the work of Delgado—not because I wanted to replicate or extend these experiments but because I found them so fascinating. I searched out neuroscientists who had met or worked with the man. They tended to describe him as an authoritative figure who spoke English with a thick Spanish accent. He commanded attention and respect. And if truth be known, they added quietly, he was something of a showboat.

But who else would pull a stunt like fighting a bull that had been implanted with electrodes?

The article in the *New York Times* described an experiment that had taken place much earlier in Cordova, Spain. Delgado had developed a system to deliver electrical impulses into an animal's brain by remote control, using a radio transmitter to communicate with a receiver attached to the animal. Motivating this piece of research was Delgado's idea of comparing the brain activity of docile breeds of bulls with that of aggressive ones. Any detectable differences might at least hypothetically explain the difference in behavior. When the receiver got a signal from the transmitter, it activated the circuit attached to one or more of the electrodes and delivered current into the brain. Delgado operated on bulls to implant electrodes and attached the receiver unit. He put the electrodes in areas of the brain related to the motor system— stimulation often invoked head-turning, leg-lifting, and vocal-izations.

In the arena, Delgado waved a red cape at a bull. When the bull charged, Delgado pressed a button on the transmitter and the bull suddenly came to a stop.

As typical with Delgado's research, he photographed as well as filmed the behavior of the animal. By capturing the behavior on film Delgado had a permanent record he could study in his lab, as well as something to show other researchers and support the conclusions he put in his reports. And a few publicity stills never hurt either.

Titled "'Matador' With a Radio Stops Wired Bull", the *New York Times* article gushed with enthusiasm, claiming the ex-periment "was probably the most spectacular demonstration ever performed of the deliberate modification of animal behavior through external control of the brain."

Other researchers weren't so sure. It's not so hard to get an animal to stop moving if you scramble the signals in the part of the brain that controls the muscles. Delgado said that he could also get the bull to turn and trot off, as if the animal decided not only to halt its charge but go somewhere else. Looking at the film, the animal looks kind of shaky to me.[2] It's not astonishing that you can get an animal to turn or move in a certain direction by giving the brain a little spurt of electricity in the right spot in one hemisphere or the other. But according to Delgado, "The result seemed to be a combination of motor effect, forcing the bull to stop and to turn to one side, plus behavioral inhibition of the aggressive drive. Upon repeated stimulation, these animals were rendered less dangerous than usual, and for a period of several minutes would tolerate the presence of investigators in the ring without launching any attack."[3] Or they might have just been confused.

Here again, I wonder what the bull was thinking. Was it something like, "Hey, what just happened?" Or maybe: "I didn't really want to charge that guy, I'm sure he was no threat."

Although the *Times* article didn't speculate on this, it mentions that Delgado believed his research "may be of decisive importance in the search for intelligent solutions to some of our present anxieties, frustrations and conflicts." Some of those solutions apparently involved modification of behavior. "For example," the article says of Delgado, "he has been able to 'play' monkeys and cats 'like little electronic toys' that yawn, hide, fight, play, mate and go to sleep on command." At the end of the article Delgado is quoted as mentioning "scientifically programmed education."

Phrases such as "little electronic toys" bandied about in a widely read newspaper frightened a lot of Delgado's colleagues. It frightened a lot of other people too.

But Delgado was determined. As is the case with most pioneers, he had to struggle for years to perfect his techniques. His wireless system was ingenious, and he could deliver stimulation to animals by either a timer, which would inject current at certain intervals, or by remote control. He often called the device a stimoceiver. Many of his earliest subjects were monkeys, who were agile and clever enough to defeat his initial efforts. As

Delgado himself admitted, the animals scored many early victories by chewing wires, transistors, and fingers. But finally he and his research team managed to monkeyproof the system as much as possible.

Delgado chose to study monkeys for several reasons. Not only is the monkey's nervous system quite similar to ours, the animal also forms groups with strong social interactions and hierarchies. Using a social animal as a test subject gave Delgado an advantage that Hess didn't have with his cats. Maybe you've watched a baboon bare its fangs on the Discovery Channel or Animal Planet and wondered who he's threatening—then you notice that none of the other baboons are the slightest bit fazed because the tooth-barer was merely yawning. Animals of the same species are able to interpret the actions of a stimulated test subject with much more fidelity than a human observer. When Delgado placed the test subject in a group, he could not only study the stimulation's effect on the animal's social behavior, he could study how other animals interpreted it as well.

* * *

For several years in the late 1950s and early 1960s Delgado kept a colony of 4-6 monkeys in an enclosure fronted by a plate of glass.[4] Monkeys were added or taken away as the experiment proceeded. The species was Macaca mulatta, also called macaques. (They're also known as rhesus monkeys, a name given to them by Audebert, a French painter, though no one is entirely sure why.) These monkeys have brown or grayish fur and pink faces. They are common in India and other parts of Asia. Adults range in size from 8 pounds all the way up to about 25 pounds, with males generally much bigger than females. In the wild they form social groups that include males and females. Both genders establish a social ranking—a boss and a set of followers of various ranks—but males are the only ones who tend to get violent about it.

Individual monkeys are usually recognizable, but Delgado made it even easier to identify them by marking symbols on their fur (he used Miss Clairol hair dye). The researchers implanted electrodes in some of the monkeys while the others were observers. Stimulation was wireless, sometimes under the control of a timer on the monkey and sometimes with a transmitter/receiver combination. The receiver weighed less than 4 ounces and the monkeys easily tolerated wearing it on their backs. The stimulator was attached to a collar and the wires leading to the skull were subcutaneous (embedded underneath the skin), otherwise the monkeys would tear them out. Different animals could be controlled separately since they wore receivers tuned to different frequencies.

The boss of the group was the alpha male, the largest and most aggressive animal in the enclosure. He usually took ownership of one corner and a considerable amount of space surrounding it, and insisted that the other monkeys respect his privacy. The rest of the group crowded against a far wall.

Delgado lumped behaviors into categories such as grooming (which monkeys do a lot of), eating, resting, etc. The behavior of the monkeys followed dominance relations — who attacked whom, who was submissive, and so forth. Delgado discovered that the level of aggression could be easily altered by stimulating certain brain regions.

When the researchers stimulated the dominant male's brain in an area known as the central gray, he became enraged and attacked the others.[5] The attacks weren't particularly fierce, and they were conducted in a normal, coordinated manner rather than just a wild lashing out. This means it was probably a whole neural system that ultimately got activated — a system that controlled a sophisticated set of aggressive behaviors — rather than just a few neurons. Even more evidence to support this comes from the target of the male's aggression: he didn't attack just anyone, he picked on monkeys that had given him trouble before.

This is interesting. It wasn't uncontrolled aggression. He didn't get mad and strike the nearest object. He used judgment in the choice of individuals to attack — either because he already had a reason not to like them, or because he was taking it out on somebody he knew had acted submissively before (sort of like getting mad when you're at the office and then going home and kicking the dog). So the electrical stimulation didn't change the monkey's assessment of the social hierarchy or cause him to lose control, it just made him grumpy.

It doesn't seem to be the same with cats. Researchers have found that a cat with stimulated rage (sham rage) will diligently run a maze to find something such as a rat to attack. And if another cat is nearby, it will do nicely as a target for the attack, even if the cats were behaving like best buddies only moments before. Imagine the surprise of the other cat when his companion suddenly starts hissing at him. That's the end of the friendly relationship—Delgado notes that friendly cats can be turned into enemies by such stimulation.[6] They don't, in other words, kiss and make up.

But cats aren't social animals like monkeys (and humans), and since they're predators, it's perhaps hard to distinguish between hunting and fighting behavior. In monkeys, the stimulated rage had a definite impact on the group. In some cases when the rage of the alpha male was stimulated, the female who tended to be his favorite (in terms of sexual activity) also attacked the others, even though she was much smaller. She assumed, perhaps, that something serious was afoot, and had a vested interest in seeing alpha hold his position in the troop.

Delgado also noticed that when beta, the second-ranked member of the group, was stimulated instead of alpha, beta sometimes gained an advantage and lifted his status to the top. Apparently the boss was intimidated. I suspect this might have occurred because of the release of male hormones in the beta animal that accompanied the stimulation of the aggression circuits. It was enough, it seems, to push a monkey that was a close second all the way to the top of the heap. For the lowest ranking members, however, such stimulation didn't help. The cellar-dwellers only managed to get beat up.

Other changes in behavior could be initiated by stimulating different sites in the brain. Stimulation of a certain part of the thalamus of a female had a peculiar effect. The thalamus is a vital nucleus. It's actually a collection of nuclei, each of which makes specific synaptic connections, often with parts of the cortex. Some of these thalamic nuclei are involved in processing sensory information such as sight or sound, and some carry motor information that help coordinate movement. One particular part of the thalamus (nucleus medialis dorsalis of the thalamus) invoked a pattern of odd movements in which the female jumped on and off the wall of the cage. Although to a human observer the activity would seem to have little meaning, one of the male monkeys evidently found it a turn-on, for he repeatedly mounted her shortly afterward.

At other stimulation sites, monkeys often made a series of unnatural movements. In some experiments Delgado set the timer on the stimulator to deliver a current at specific intervals such as one minute. Often he observed the monkey perform the same pattern of movements independent of whatever it was doing at the time.

The monkeys noticed the awkward movements too. Once more I have to wonder what they thought of it. Of particular interest were the repeated, periodic stimulations, which could not be explained away as just some kind of muscle twitch (which is what I imagine I'd think it was, in their place). It seems that in certain cases the monkeys became so aware of the stimulation that they could time it. Delgado reported that some monkeys anticipated the next stimulation by bracing themselves to minimize the disruption of their activities.

This fascinates me. The activation of this unwilled movement came from the monkey's own brain—albeit from an injected current—and other parts of the brain recognized the situation well enough to prepare for it. Would these parts have learned to stop the movement over time instead of just bracing for it? Which part of the brain can learn to control what part of the body (and mind)? How independent are the parts of the brain and mind? I wish I knew the answers to these questions.

As with the experiments of Hess, you can ask whether stimulation of the monkey's brain activated normal mechanisms.

In other words, are these the same neural circuits that would get activated when the monkey chooses to perform these same activities (as opposed to the experimental situation in which the activities are elicited involuntarily by electrical stimulation)? Delgado believed the answer was yes. As evidence he cited the fact that stimulation often invoked simple movements, yet the act of making these movements required the assistance of a number of supporting systems. For example, if the monkey stops and scratches his back, the animal's nervous system must correctly activate postural and other muscular systems to keep from toppling over. Since the electrical stimulation couldn't have directly activated such a dispersed set of circuits, the neurons responsible for these support functions must have been recruited as they would have been for a normal voluntary or willed back-scratching. As additional evidence, Delgado noted that stimulated acts of aggression were calculated rather than random.

Delgado also discovered that he could switch aggression off as easily as he could switch it on. If he stimulated a part of the caudate nucleus (a large member of a group of important nuclei not far below the surface of the brain), an aggressive alpha male could be transformed into a docile animal. Delgado used this trick when he needed to move or handle the normally uncooperative animal. The other monkeys in the troop noticed it too. Within 6-8 minutes the other monkeys lost their fear of him, invading his space with impunity. Yet Delgado reported that after stimulation the only visible change in the alpha male was a slight relaxation of the body. But that seemed to be enough. Stance and posture are apparently critical in monkey society, as is true with humans as well.[7]

Something even more interesting happened when the researchers placed a lever in the cage and rigged it to activate alpha's caudate nucleus. This was spicy enough for Delgado to get it published in the prestigious journal *Science*.

The *Science* paper describes a colony of four monkeys, two males and two females: M1, M2, F1, and F2, with M1 and F1 the dominant male and female.[8] M1 was a fearsome, intimidating alpha male who had a propensity to chew on his own hand. Delgado placed the lever in the cage near the food and all animals had equal access. The researchers implanted electrodes in M1 and F1 but not M2 or F2.

When the lever was configured to stimulate M1's caudate nucleus, pressing it would greatly reduce M1's aggression. Any animal could press the lever. If M1 did it, this would in effect be self-stimulation. If another animal pressed the lever, Delgado called it heterostimulation (employing the Greek-derived suffix *hetero*, meaning other or different), since the actions of this animal would be stimulating another.

Monkeys, being curious creatures, investigated the lever and occasionally pressed it. M1 did so on occasion, but seemed to derive no satisfaction from it. F2, however, began pressing the lever many more times than the others. Delgado argued that she learned the trick—when she pressed the lever, the bully M1 went limp for a while. To prove his argument, Delgado not only pointed out the higher number of times F2 pressed the lever, he also claimed that F2 did it at least occasionally after M1 gave her threatening glances. In the *Science* paper Delgado published a picture of F2 pressing the lever while looking at M1. While it doesn't prove Delgado's assertion that she knew what she was doing, it's pretty convincing. Although M1 isn't doing anything particularly menacing (at least not at the precise moment the picture was taken), subordinate monkeys rarely stare at the dominant monkey—in monkey etiquette this is considered a threat or challenge, and would probably not go unanswered. I believe she knew the trick. And besides, Delgado filmed his experiment, so I have to assume he could back up whatever claims he made in *Science*.

Delgado also configured the lever to stimulate F1. None of the monkeys including F1 showed any special inclination to press it. When the lever was configured to stimulate a part of M1's brain that increased his aggression, pressing the lever became an even more unpopular thing to do. None of the monkeys activated the switch more than a few times.

Delgado ended the *Science* article with an understated reference to the possibilities of brain stimulation: "Hetero-stimulation presents obvious questions about hierarchical control, reciprocal punishment, instrumental self-defense, and other problems related to human behavior."[9] But somebody in the government was evidently interested in those things. Funding sources for Delgado's monkey experiments included grants from the U.S. Public Health Service and the Office of Naval Research.

\* \* \*

While Delgado was wiring up his monkeys and Olds was watching rats stimulating themselves to exhaustion, a group of physicians at the Tulane University School of Medicine was exploring electrical stimulation of the brain as a therapy for psychiatric patients.

Even before Olds discovered that stimulation of certain sites such as the septal region was reinforcing in rats, a team led by Robert G. Heath, chairman of the Department of Psychiatry and Neurology at Tulane, discovered that stimulation of certain subcortical areas made human patients calmer. They made this discovery in conjunction with research on last-ditch therapies for treatment of schizophrenia and other debilitating disorders.

Before I describe the reports published by Heath and his colleagues, I want to emphasize several points. Although it's possible that the results of these therapy studies are applicable to humans in general, it's not at all certain. These patients were seriously incapacitated because of their illness, which of course was the reason for their participation in the experimental treatments. Because of their emotional turmoil, it is likely that the parts of the brain that process emotion were severely affected. The subjective responses of these patients to electrical stimulation of the brain may or may not be similar to people who are not suffering from these disorders.

Another important consideration is the health of the patient. These procedures were drastic and involved plenty of risks, including infection, seizure induction, and brain damage caused by the insertion of deep electrodes. The motivation underlying the pursuit of this form of treatment was a rather weak theory—and not a widely accepted one—that specific subcortical structures and/or their connections to the cortex were involved in development of schizophrenia. But this was a time of desperation in the treatment of mental illnesses such as schizophrenia. In the 1950s and early 60s, when most of this work was carried out, hospitals were overflowing with patients who didn't respond to any form of treatment. In the latter half of the 1950s conditions began to improve and the crowding was reduced with the development of effective drug treatments, such as chlorpromazine for schizophrenia. Prior to this, however, physicians and patients would try almost anything.

Perhaps mindful of what happened to Roberts Bartholow, Heath and his colleagues stressed their goal in this work was medical, not scientific. "In the series of depth electrode studies at Tulane," they wrote in a paper published in 1963, "the primary motivation has always been therapeutic. Only patients who have failed to respond satisfactorily to existing therapies have been studied and treated with these techniques."[10] Even so, they eventually ran into trouble, as did Bartholow, though in the beginning their work was not severely criticized and indeed was published in high-quality journals. More on the ethics issue later in the chapter.

Citing the work of Olds, Heath and his colleagues published several papers in 1963 on self-stimulation in humans. The results weren't quite what one would expect after hearing so much about "pleasure centers" in the brain.

One report, published in *Science*, described studies performed on a patient who had been diagnosed with catatonic schizophrenia. Such patients tend to spend a great deal of time rigid and motionless, sometimes in contorted postures. At other times they express highly disordered thinking and exhibit emotional outbursts unrelated to their surroundings or the people around them. The physicians implanted electrodes in several regions, including the head of the caudate nucleus, septal region, amygdala (a nucleus in the temporal lobe known to be involved in processing emotions), parts of the thalamus and hypothalamus, and a collection of cells deep in the base of the brain called the tegmentum.

A major problem confounding the results of this experiment was the patient's perseveration—the tendency to continue a behavior much longer than required or appropriate. When presented a switch containing the button for self-stimulation, the patient pushed it and often kept on doing so, long after the current had been quietly turned off.

You get the impression this fellow was in extremely bad shape. Heath didn't even bother getting the patient's subjective impressions of the brain stimulation. Instead, stimulations were operationally classified as rewarding or aversive, based on the subject's rate of response. But due to the patient's perseveration, Heath had to get creative in the experimental design. Sometimes he discreetly changed the setting so that a button controlled different electrodes, and sometimes he made the patient test each of a set of buttons before deciding which one to continue pushing.

The failure to consider the patient's reported experiences makes this study a lot less interesting to me. It's rather the same disadvantage as using laboratory animals—they can't tell you what they're feeling, so you have to go by what they do rather than what they say.

The patient found some stimulations reinforcing and some aversive. In one trial the patient preferred to stimulate his brain rather than eat, even though a food tray was placed nearby and the patient must have been hungry because he hadn't eaten in 7 hours (Heath says he "ate heartily" after the trial). Other stimulations were avoided or repeated much less often. But Heath discovered that the most important factor that determined whether stimulation of a particular site was reinforcing or aversive was the level of current rather than placement of the electrode. Stimulation became aversive when the current was increased 25-100 percent above the reinforcing level at any given site.

For subjective reports we have to turn to a paper published in the *American Journal of Psychiatry*. This longer article describes two self-stimulation patients whose conditions seem much better than the person with catatonic schizophrenia, although of course these patients still had serious enough problems to warrant a risky and experimental procedure.

A patient identified as B-7 by Heath had been diagnosed with severe narcolepsy—a sleeping disorder in which the patient often suddenly and unwilling falls asleep—and cataplexy, which refers to a sudden loss of muscle tone. Heath and his colleagues implanted electrodes and the tests began six months after surgery. After some preliminary testing, the physicians configured a stimulator with three buttons, each controlling an electrode at one of three sites: septal region, hippocampus, and tegmentum. The patient wore the stimulator control on his belt for 17 weeks. He could press any button at will, but a timer prevented continuous stimulation.

The septal region proved most popular, but not for the reason you might think. According to Heath, "The patient, in explaining why he pressed the septal button with such frequency, stated that the feeling was 'good'; it was as if he were building up to a sexual orgasm. He reported that he was unable to achieve the orgastic end point, however, explaining that his frequent, sometimes frantic, pushing of the button was an attempt to reach the end point. This futile effort was frustrating at times and described by him on these occasions as a 'nervous feeling.'"[11]

I wonder if the animals in the Olds and Milner experiments would have said something similar. It's like being almost but not quite in paradise—knocking on the door, so close you can see it, hear it, smell it. And then, disappointment.

The septal stimulation also had the great benefit of keeping Heath's patient awake and alert, which circumvented his narcolepsy. His friends even helped him out: on occasion the patient fell asleep so rapidly he didn't have time to press the button, so somebody did it for him. Heath reported that the patient's symptoms were so improved "he was employed part-time, while wearing the unit [the stimulator], as an entertainer in a night club."[12] Must have been some act.

Stimulation of the hippocampus in this patient was "mildly rewarding." The tegmentum was a different story. Although when the patient pressed this button he felt alert, which was the main goal, he also complained of "intense discomfort." Apparently the stimulation was so obnoxious that the patient prevented the button from being accidentally pressed by jamming a hair pin under it.

The other patient, B-10, suffered from epilepsy along with emotional disturbances that resulted in violent outbursts. His regimen was a little more complicated. He had electrodes in many different regions, including the centromedian thalamus, septal region, caudate, tegmentum, amygdala, hippocampus, paraolfactory area (another one of those areas involved in processing emotion) as well as four different cortical areas: parietal, frontal, occipital, and temporal.

In one phase of the testing, the patient was provided the opportunity of stimulating each region for a small period of time. The one he stimulated at the highest rate was the centromedian thalamus—nearly 490 stimulations per hour. Yet he didn't report the sensation as pleasurable. Remarkably, it left him frustrated, similar to B-7's stimulation of the septal region, except in B-10's case he wasn't chasing an "orgastic end point" but rather a fleeting memory that for some reason he really wanted to experience. He became angry and frustrated.

The septal region, however, was described as pleasant—"feel great." He also liked the tegmentum, which he said gave him a "drunk feeling" and called it a "happy button" (in contrast to the previous patient, who hated it so much he disabled that button). Amygdala stimulation was mildly rewarding but disappointing— the patient requested an increase in current.

In contrast, stimulation of the hippocampus made the patient "feel sick all over." He also didn't like to press the paraolfactory button. None of the cortical sites gave this patient any significant feeling, and he rarely pressed these buttons when given the opportunity.

During another phase of the testing, B-10 was given a three-button stimulator like the one used by B-7, though the period of the test lasted only 6 hours rather than for weeks. The buttons controlled the three electrodes he had used frequently in the earlier phase: centromedian thalamus, septal region, and tegmentum. The patient began stimulating the centromedian thalamus at the highest rate but after a couple of hours became thoroughly frustrated. To alleviate the irritation he would press the other buttons, and by the end of the six hours he was pressing the septal button at a slightly higher rate than the centromedian thalamus.

Heath also described a dramatic test of the efficacy of the septal stimulation to calm the patient during one of his violent episodes. This wasn't part of the self-stimulation phases—an observer activated the stimulation without the patient's knowledge. The change in behavior according to Heath was instant and remarkable, replacing the rage and anger with peace and contentment. The patient said he felt the onset of sexual arousal but couldn't explain why his mood changed so abruptly. (The sexual feelings sound more like what B-7 described with his septal stimulation, though they weren't part of B-10's description during self-stimulation—or at least Heath didn't include this in the paper.) B-10 said he didn't know why he started thinking about sex, it just popped into his mind.

* * *

Delgado was also involved in the testing of the efficacy of brain stimulation in human patients. He believed the treatment, though drastic and reserved for intractable cases, offered a reasonably safe alternative to draconian measures such as removal of brain tissue. In 1967 he wrote, "the use of electrodes represents a more conservative approach than the destruction of portions of the brain, which may now be regarded as necessary in the treatment of special cases of pain or involuntary movements. In some cases, long-term electrical stimulations may obviate the need irrevocably to destroy cerebral tissue by surgical operation."[13] As long as you're going to do the craniotomy anyway, why not try stimulation? Delgado went on to bring up a point raised by a lot of advocates of electrical stimulation of the brain: we do it for the heart (in pacemakers), why not the brain?

The brain stimulations carried out by Delgado and his colleagues evoked the gamut of human emotions, from violent to passive. Anger was clearly evident in some patients, particularly those predisposed to violence. One patient was a young woman who had suffered from encephalitis in early childhood and experienced temporal lobe seizures throughout her life. Her behavior could unpredictably swing to violent extremes and she was institutionalized after committing numerous assaults. Physicians were able to replicate these outbursts when they stimulated the amygdala in her right hemisphere. One moment she would be playing a guitar and singing sweetly, then, after stimulation, she threw the instrument away and went into a rage. Once the physicians narrowed down the area of the brain in which this behavior could be evoked, it was destroyed. In this instance it was better to take out the diseased tissue, but to Delgado's point, the stimulation helped physicians find and remove the smallest amount.

Pleasure and giddiness could also be evoked. One patient, a 36-year-old woman with intractable epilepsy, had a startling change in behavior after temporal lobe stimulation: "She started giggling and making funny comments, stating that she enjoyed the sensation 'very much.' Repetition of these stimulations made the patient more communicative and flirtatious, and she ended by openly expressing her desire to marry the therapist."[14] When she wasn't being stimulated "her behavior was quite proper, without familiarity or excessive friendliness." Stimulation at other sites had little effect on her mood.

In my opinion, the most intriguing phenomena Delgado reported came from patients who seemed to attribute the effects of brain stimulation to their own volition. Stimulation of one patient caused him to turn his head and lean over, as if in search for something. This was a reliable effect, repeated several times. "The interesting fact was that the patient considered the evoked activity spontaneous and always offered a reasonable explanation for it. When asked 'What are you doing?' the answers were, 'I am looking for my slippers,' 'I heard a noise,' 'I am restless,' and 'I was looking under the bed.'"[15]

What about the long-term effects of electrode implantation? Delgado didn't see a problem. "Leaving wires inside of a thinking brain may appear unpleasant or dangerous, but actually many patients who have undergone this experience have not been concerned about the fact of being wired, nor have they felt any discomfort due to the presence of conductors in their heads. Some women have shown their feminine adaptability to circumstances by wearing attractive hats or wigs to conceal their electrical headgear, and many people have been able to enjoy a normal life as outpatients, returning to the clinic periodically for examination and stimulation."[16]

Delgado summarized his animal experiments and therapeutic testing in 1969 with the publication of a book called *Physical Control of the Mind: Toward a Psychocivilized Society*. I've quoted from this book several times above, and it's a fascinating book, well worth reading.[17]

The book is not only fascinating, it's rife with what seem to be contradictory sentiments. Delgado worried over the future of a world in which people wielded nuclear weapons capable of global destruction. This, remember, was a time of heightened tensions — Vietnam, the space race, and the ongoing cold war — and not long after Kennedy played nuclear chicken with Khrushchev during the Cuban missile crisis. Delgado began looking for science to solve the problems caused by human aggression. "I have been concerned by the urgent need to resolve archaic human antagonisms," he wrote, "and to find a new conviviality based on biological reality rather than on wishful thinking."

The title is provocative and calls to mind the "little electronic toys" comment. At one point in the book Delgado asks if the mind can be physically controlled. Throughout the text he seems to waffle on the answer.

At one point he suggests that electrical stimulation of the

brain, which he abbreviates ESB, is just a research tool rather than the solution to our problems: "Possible solutions to undesirable aggression obviously will not be found in the use of ESB. This is only a methodology for investigation of the problem and acquisition of necessary information about the brain mechanisms involved." But then you'll find statements like this: "The old dream of an individual overpowering the strength of a dictator by remote control has been fulfilled, at least in our monkey colonies...."

Although Delgado wrote, "Human relations are not going to be governed by electrodes," he also wrote, "It is clear that manifestations as important as aggressive responses depend not only on environmental circumstances but also on their interpretation by the central nervous system where they can be enhanced or totally inhibited by manipulating the reactivity of specific intracerebral structures." Elsewhere he wrote of ESB being a "master control of human behavior" and, "We know that by electrical stimulation of specific cerebral structures we can make a person friendlier or influence his train of thought."

But he mentioned limitations. "Contrary to the stories of science fiction writers, we cannot modify political ideology, past history, or national loyalties by electrical tickling of some secret areas of the brain." And yet, "The individual is defenseless against direct manipulation of the brain because he is deprived of his most intimate mechanisms of biological reactivity. In experiments, electrical stimulation of appropriate intensity always prevailed over free will; and, for example, flexion of the hand evoked by stimulation of the motor cortex cannot be voluntarily avoided."

If I'm interpreting Delgado correctly, he felt that, in contrast to simple movements such as the flexing of a hand, brain stimulation couldn't bring about a complex thought or emotion unless it was already present. "By means of ESB we cannot substitute one personality for another, nor can we make a behaving robot of a human being. It is true that we can influence emotional reactivity and perhaps make a patient more aggressive or amorous, but in each case the details of behavioral expression are related to an individual history which cannot be created by ESB." Delgado called the notion of a dictator controlling people with brain stimulation an "Orwellian possibility," referring to the dystopic novel *1984*, "but fortunately it is beyond the theoretical and practical limits of ESB." Delgado believed drugs or indoctrination would be far more effective in directing behavior.

Delgado summed up the goal of his research with this sentence: "The main aim is to establish a scientific foundation for the creation of a future psychocivilized society based on a better understanding of mental activities, on the liberation from the domination of irrational mental determinants, on the estab-lishment of personal freedom through intelligent choice, and on a balance between the material and mental development of civili-zation with the power of reason directing the use of increasing technological control of our physical environment." Exactly how we liberate ourselves from "irrational mental determinants" is an interesting question, and the "increasing technological control of our physical environment" sounds a bit ominous.

Interestingly, Delgado claimed he got letters written by "many people" who volunteered to have electrodes implanted in their brains. Some of them, he noted, were apparently motivated by "pure scientific interest" while others exhibited signs of "psy-chotic disturbances." Delgado turned them all down.

* * *

The hounds started baying by the time the 1970s arrived. Lawsuits over the human brain stimulations threatened to end the work of Delgado and his colleagues, and in 1974 Delgado left Yale. He repeatedly stated he left on his own accord after receiving a tempting offer to help build a new medical school in Spain.

Heath was in for a rough ride as well. He and his colleagues performed brain stimulation on about a hundred patients from 1950 through the early 1970s, and though in the beginning the procedures did not always go well, most of the public discussion of their work was laudatory.[18] They achieved some success with these experimental treatments in the first few years, but during this time the electrode placement wasn't carefully done and it is possible that the improvement seen in some of the patients was due to selective and accidental damage to their brains rather than electrical stimulation. (Treatments for severe disorders often include deliberately inflicted damage to brain tissue, as I've mentioned in this book: surgical removal or the application of heat or high-current electricity to burn up the trouble spot.) When the physicians began employing stereotaxic procedures, the effectiveness of stimulation therapy strangely declined.

With the development of effective drugs to treat schizophrenia, experimental therapies became harder to justify. But Heath continued to claim that pleasurable brain stimulation aided certain patients, and suggested it be tried as a novel approach in treating drug addicts. But his work on the pleasurable aspects of human brain stimulation has not been widely regarded in the scientific community. A 1993 review of subjective experiences that have been invoked by human brain stimulation said the pleasurable response to septal stimulation found by Heath and Delgado "has not been confirmed."[19]

As criticisms began to mount, Heath didn't help himself when he published a silly paper in 1972. In the course of stimulating a homosexual patient who had a variety of psychiatric problems including suicidal thoughts, paranoia, and drug abuse, Heath decided to see if he could change the patient's sexual orientation. With the help of a young woman, Heath and his colleagues attempted to condition the patient to the joys of heterosexual experience by combining it with septal stimulation, which was assumed to act as a reinforcer.[20] The patient subsequently reported bisexual activity. I suspect Delgado would have said the patient was probably bisexual to begin with.

The sexual reorientation experiment raised eyebrows. So did the occasions when the physicians, acting as scientists, elicited angry or violent responses in patients. Critical articles began to appear. Peter Breggin, who has often found fault with psychiatric treatments, ranted specifically against the Tulane group in the 1970s.

Heath defended himself in a manner similar to Roberts Bartholow. For instance, Heath argued he obtained proper consent from all the patients, though this is problematic considering their mental state. In 1996 Heath published a book, *Exploring the Mind-Brain Relationship*, defending his work and describing the experiments.[21]

It's interesting to contrast Heath with Penfield. One was widely admired, the other not so much. Penfield's patients seemed to have had better outcomes in general, so that's an important point. But I would think the studies of Heath (as well as Delgado) were far more interesting than Penfield's work. Yet Penfield's research remains fairly well known, while Heath has faded into obscurity. He passed away in 1999.

It is also interesting to contrast Heath and Delgado with James Olds. While Heath and Delgado lived on the edge, Olds was much more cautious. And Olds seemed to become even more cautious as Delgado's fame and notoriety grew. I wonder if Olds looked at Delgado and started thinking to himself, What have I begun?

After Delgado left Yale, he drifted away from the work he had spent decades mastering. Instead of using invasive methods of brain stimulation, Delgado pursued noninvasive techniques that don't require a craniotomy, such as exposing a patient's head to electric fields. But he continued to decry the prejudice against electrical stimulation of the brain. In an article published in 1981 he wrote that acceptance of brain stimulation as a means of therapy faces stiff opposition from religious and emotional arguments. "Science fiction also has played a harmful role by inventing radio-controlled individuals and robotized armies."[22] This from a man who famously stopped a charging bull by stimulating its brain, and who titled his book *Physical Control of the Mind*.

Delgado passed away in 2011. If people at Yale had any hard feelings about his departure or any misgivings about his work while at the university, they didn't show it in the press release the Department of Psychiatry issued on 28 November 2011 to mark his passing. Calling him "a groundbreaking pioneer in brain-stimulation research", Yale noted that his work "provided concrete evidence that neural activity produced complex behaviors." (There was a doubt about this in the 1950s and 60s? I don't think so.) The press release also referenced the modern revival of electrical stimulation, saying Delgado's "work paved the way for the development of deep brain stimulation treatments for psychiatric disorders, including depression and obsessive-compulsive disorders."[23] Much more on this later.

I believe Delgado made a good point about sensationalist attacks on an incipient and emotionally charged subject such as electrical stimulation of the brain. It's proper to question anything new, especially something potentially dangerous, but harangues are rarely helpful, and distortions never are. But there's always somebody to hit the panic button—and profit from newspapers, magazines, books, and speaking arrangements in the process. When there are rabble to be roused and a buck to be made, there will always be skilled writers and orators ready to don a halo and express their righteous indignation.

Think of how many millions of lives antibiotics and vaccinations have saved. But testing for these and other treatments relied on questionable ethics—not necessarily questionable in their day, perhaps, but questionable by today's ethical standards.[24] Were the experiments criminal, or would it have instead been criminal to forego the treatments and lose millions of lives in the future?

There's a line to be drawn somewhere. A few people think it's criminal that biologists today still experiment on our furry friends. At the other end of the spectrum are people who have no qualms about using humans for experimental purposes—for instance, atrocities have been committed by people in Hitler's Germany, in militarized Japan of the 1930s and 40s, and also in the United States during the 1950s, when "intelligence" agencies conducted mind-control experiments.[25]

A reasonable middle lurks somewhere in between these extremes, though I doubt we'll be able to find it in this 24/7 news cycle where "news" seems to be devoted mostly to contests of who can express the most outrage—or be the most outrageous. News programs of today remind me a lot of professional wrestling. I think they use the same business model.

By the middle of the 1970s, the go-go era of science was over. Budget constraints and ethics concerns killed it off. Researchers began turning to more attainable, less quixotic goals, such as the subject of the next chapter.

# 7
# Sensory Prostheses: Seeing and Hearing with Neuroelectricity

I've avoided a strict definition of what exactly falls within the category of electrical stimulation of the brain. In this chapter most of the techniques I will describe don't initially stimulate the brain, if by "brain" you're referring only to the cerebral hemispheres, cerebellum (the "little brain" that's tucked underneath the occipital lobe in the back), and that thin piece sticking down called the brainstem that connects with the spinal cord.

How do you get at the brain without drilling holes in the skull? You can do it transcranially—across the skull—if your stimulation technique involves radiation or perhaps an electric or magnetic field that can manage to penetrate the bones surrounding the brain. Or you can get sneaky about it and slip in through one of the nerves.

Recall that nerves are bundles of axons—nerve fibers— snaking through the body. An axon is the business end of a neuron—a lengthy extension of the cell along which the action potential travels. Action potentials carry information, often in their rate or frequency, and also in their timing, particularly in relation to the action potentials of other neurons.

Some nerves carry sensory information. This information comes from sensors embedded in the periphery, such as in the eye or ear or skin, which transduce energy in the form of light, sound, or pressure into an electrical potential.[1] The electrical potentials get transformed into a train of action potentials, typically at a rate commensurate with the intensity of the stimulus—a bright light produces more action potentials than a dim one, for example. At some point these action potentials are carried to the brain via axons of neurons that form the sensory nerves.

Information in motor nerves goes the other way, from brain to periphery. These nerves arise from motor neurons in the brain and spinal cord. Action potentials in these nerves act as signals to activate muscles (which in turn produce action potentials of their own, instigating contraction).

So the central nervous system—the brain and spinal cord— behaves as sort of a central processing unit, or CPU, of the body. Data comes in along sensory nerves, and after some sort of computation or processing occurs in the brain, the motor nerves carry the output to the muscles to execute whatever behavior the brain has decided upon.

Occasionally something goes wrong with this flow of information. An interruption in the motor pathway causes paralysis. Interruption of the sensory pathway causes a loss of sensation. Deafness, for example, may arise from the death of receptors that transduce sound into electrical activity, or from destruction of the neurons whose axons form the auditory nerve, or a break in the auditory nerve, or from a disease that attacks the brain regions which process auditory information.

In the types of deafness caused by receptor loss—which is unfortunately a rather common problem—biomedical researchers in the 1970s began to wonder if they could build a system that might be able to turn sound into electricity and pipe it into the brain, bypassing the missing receptors. The go-go era of science was pretty much over, but scientists finally decided to get earnest about an ingenious way to transport sound to the brain of the hearing impaired. People would be able to "hear" with electricity.

It's a roundabout way of stimulating the brain. Ultimately you're doing something similar to the work of Penfield and his colleagues, who elicited sounds and other sensations by stim- ulating the cortex responsible for processing those sensations, but you're doing it by way of a sensory nerve.

Nerves are the backdoor to the brain. Some viruses such as herpes simplex and rabies invade the brain, possibly via nerve pathways. With a nerve you can have access to the brain without having to open up the skull.

This chapter focuses on prosthetics—the use of devices to replace missing or impaired body parts. Sensory prostheses have been developed to compensate for hearing loss, and visual system prostheses are presently under intense development. I won't cover the other senses, which have received far less attention, although there has been some effort to develop an olfactory (smell) prosthesis.

* * *

The ear is a marvelous transducer. It seems a little like a Rube Goldberg[2] contraption with all its complexity, but it's actually an efficient mechanism to turn sound energy into electrochemical signals—the language of the brain. The complexity arises in part because it's an evolutionary development that has seen quite a few adaptations over the years.

The outer ear, called the pinna, funnels sound into the ear canal, the end of which is a taut membrane called the eardrum. Sound waves set the eardrum into motion. This motion initiates a vibration in a set of three tiny bones in the middle ear called the ossicles—the malleus is connected to the eardrum, the incus is connected to the malleus, and the stapes is connected to the incus. The other end of the stapes butts up against the cochlea, a small fluid-filled canal in the inner ear that is wound up about two and three-quarters times and looks like the shell of a snail. (The word *cochlea* derives from a Greek word meaning snail.) In humans the cochlea is about 1 centimeter (0.4 inches) in diameter, but because it is coiled up the canal is about 3.5 centimeters (1.4 inches) in length.

Little receptors called hair cells are anchored to a membrane in the cochlea called the basilar membrane. These cells have "hair" in their name because they have tiny filaments called cilia sticking out of the end that isn't anchored. The cilia brush up against a rigid membrane called the tectorial membrane. The hammering of the stapes causes the cochlear fluid to move, which produces a relative movement of the membranes of the cochlea. As a result, the flexible cilia start to bend. The bending opens mechanically sensitive ion channels and current flows through the receptor's membrane, changing the cell's potential. This is the signal—the receptors have sensed a sound wave. The receptors communicate with other cells by releasing neurotransmitters across synaptic junctions (in the usual manner that neurons communicate with one another), and the signal becomes a train of action potentials encoding the auditory information. A nerve called the cochlear nerve, which is part of the auditory nerve—also known as the 8th cranial nerve—carries the information from the cochlea to the brain, where it reaches the auditory cortex in the temporal lobe after visiting a processing center in the thalamus. The cochlear nerve in humans contains about 30,000 axons which service about 20,000-25,000 hair cells.[3]

Rube Goldbergesque, yes? But the whole thing is necessary to convert air motion into fluid motion and then to membrane motion. The air-to-fluid conversion is necessary because we live in the atmosphere above the sea, yet our cells are surrounded by salty water. The ossicles provide amplification of the vibration, which is essential because normally sound in air does not penetrate water—most sounds are reflected at the surface of water. (Which makes sense because water is so much heavier than air. Vibrations need to be much stronger to push around water than air.) Our auditory apparatus is an extremely efficient adaptation to life above sea level.

It's important to note that any sound can be broken into a spectrum of component frequencies. Your voice, for example, consists of a mixture of frequencies, with one of the main components being around 100 hertz (cycles per second) on average if you're a man and 200 hertz for a woman. Machines called frequency or spectrum analyzers can separate a sound into the individual frequencies that compose it.

The cochlea does something similar. The structure and other properties of the cochlea produce a kind of frequency analysis because a certain frequency causes movement in a certain region of the cochlea. The frequencies are arranged in order, with the highest frequencies causing movements at one end of the cochlear membranes and lowest frequencies at the other end. This is called a tonotopic map. The mapping is how the brain analyzes sound. A signal from a specific part of the cochlea means that a sound of the corresponding frequency has been detected.

This amazing feature of hearing makes cochlear implants feasible. But nobody knew about this for a long time. It took the brilliant Hungarian scientist Georg von Békésy to figure it out. Békésy went around testing cochleas from animals and human cadavers in the 1940s and 50s until he understood how they worked.

A device such as a hearing aid simply amplifies the sound electronically, but if the path of information is broken, nothing will get through. The hair cells of the inner ear are not very robust and tend to die easily. Various diseases kill them off, as does exposure to long periods of high intensity sound.[4] When they're gone they don't come back.

A number of physicians and researchers worked on early versions of cochlear implants in the 1960s and 70s. The first devices had only a single electrode, or perhaps just a few, so it stimulated the auditory nerve in just one region. Patients could hear noises, which was an improvement, but distinguishing complex sounds such as speech proved nearly impossible.

There were, of course, many skeptics in the early days of research. But when skeptics saw that profoundly deaf people who had received implants could manage to tap their feet to the rhythm of a musical recording, it was hard to deny the merits of this technique. In 1977 the United States Patent and Trademark Office issued a patent to Adam Kissiah that forms the blueprint for most of today's implants. Kissiah, an employee of the National Aeronautics and Space Administration (NASA), was motivated to develop this application because of his own hearing loss.

Cochlear implants consist of a microphone to pick up sounds and convert them into electrical signals, a processor to separate the frequencies, a stimulator, and a set of electrodes inserted into the cochlea to stimulate the nerve. Most patients receive an implant in only one ear. The microphone and processor are worn outside, around the ear. Signals are usually passed from processor to stimulator across the skin with an electromagnetic coil, which is about the size of a silver dollar. It works by induction, transmitting the signals to a receiver buried underneath the skin. Power for the stimulator's operation is also transmitted in this manner. As an alternative, the signals and power can be transferred directly from processor to stimulator with a conductor that passes through the skin, but this raises the risk of infection.

The processor serves as a frequency analyzer. The electrodes are inserted at different depths and receive signals of a certain frequency range. If the electrode is properly positioned, it will stimulate the nerve at the point along the cochlea which corresponds, roughly, to the area that would have been naturally stimulated by this frequency. Obviously if you only have one electrode you don't need the processor, but all devices these days have more than one. Most users have between 12 and 22 electrodes inserted into the cochlea, in addition to any grounding electrodes.

The trick is in proper placement of the electrodes and in processing the frequencies in a way that makes the most sense to the brain. Most cochlear implants don't go all the way to the 3.5-cm end, they stop about 75% of the way in.[5]

Unfortunately, different brains analyze sounds in a slightly different way, so one size will not fit all. And if your auditory nerve isn't functional then it isn't going to work at all.

In the United States, the Food and Drug Administration (FDA) oversees medical devices such as cochlear implants. FDA approval is required before manufacturers can market their devices in the United States. In 1984 the FDA approved the first cochlear implant for adults, and 5 years later it approved a device for children of 2 years of age or older. The minimum age has been steadily decreasing over the years. It's now 12 months. In certain cases it can be even younger, and in some countries cochlear implants have been done as early as a few months after birth.

Why so young? In general, the younger the better. The brain goes through various periods, known as critical periods, during which skills optimally develop. If a child is deprived of the opportunity to learn in this critical window, subsequent development may never be up to speed.

There's another thing you might wonder about implanting something in a child so young—won't they outgrow their implants? The answer is usually no, depending on the age. The cochlea doesn't grow much after birth. The skull does, but not as much as you might think. There are some design issues for implants intended for extremely young children, but it's not often a major problem.

According to the FDA, 219,000 patients worldwide have received implants as of December 2010. This total includes 42,600 adults and 28,400 children in America.

Millions more people could benefit, but cochlear implants don't come cheap. As of 2012 the typical cost is at least $50,000 and sometimes much more. Insurance coverage is improving but still a little spotty.

Patient outcomes have been variable. Some achieve dramatic improvement, including the ability to understand speech over the telephone and in other situations where reading lips is impossible. Even so, what they hear isn't natural sounding speech—many cochlear implant recipients describe the voices they hear as cartoonish or synthetic, at least at first. Recipients may also be able to appreciate music, although with only one or two dozen electrodes you're not going to be able to hear the richness of a symphony orchestra. Other patients aren't helped quite as much, but most gain some increased ability to understand speech. A few patients derive little benefit, unfortunately. Another problem is that the implantation reduces or eliminates the normal hearing process, so a person with a cochlear implant must rely solely on the implant for hearing in that ear. People who still have some hearing capability are therefore not generally good candidates for cochlear implants.

Language is the main thing that adults want to recover and children need to learn. A lot of research has been done and is still being done on strategies to optimize the methods processors use to parse the signals. Since the processor is external, it is accessible for fine-tuning and updating.

Language and culture are also at the root of what some people call a "controversy" associated with cochlear implants. Some deaf people have complained that this technology discourages adoption of deaf culture and sign language, and fosters a perception of inferiority toward those who do so. A few outspoken critics have called cochlear implants a genocidal attack on hearing disabled culture.[6] Outlandish remarks aside, the focal point of the debate— and in my opinion the only reasonable part of the controversy—is the extent to which hearing-impaired children should be driven to achieve "normality." It's a decision in my view best left up to the parents. The only truly unfortunate aspects of cochlear implants are the cost, which prohibits more widespread use, and the variability in patient outcome.

But it seems there are illusions and misperceptions about cochlear implants of an even greater absurdity—or at least one manufacturer of cochlear implants believes so. Advanced Bionics (a division of Boston Scientific), one of the makers of these devices, saw the need to include this bit of information in a frequently-asked-questions brochure dated November 2006: "The cochlear implant only stimulates the hearing nerve to give you access to sound. Your body does not act like a robot when you use a cochlear implant."

Robot? *Robot?* I assume the company actually believes that some prospective users are worried about cochlear implants turning them into robots. Why else would the company include a frank denial in a FAQ?

Hilarious! Maybe Delgado had a point about the extremes which some people regard electrical stimulation of the brain (see previous chapter).

For those people who have damage to the auditory nerve that precludes the use of cochlear implants, an alternative is the use of brainstem implants. These devices stimulate the auditory system in a later processing stage in the brainstem. They are trickier than cochlear implants but have been successful.

\* \* \*

Blindness can cause even more hardship than deafness. Millions of people worldwide have lost their sight. Although many of these people have rich and productive lives, the loss of vision is keenly felt. It is the most important sense, at least as judged by the amount of brain tissue devoted to it. Primates have more than 30 regions in the cortex that are responsible for the main aspects of vision.[7] At least half of the cortex in the human brain probably contributes in some way to the processing of visual information.

The first stage of vision occurs when the cornea—the outer part of the eye—and the lens bend incoming light to form an image on the retina, which is located at the back of the eye. The retina consists of several layers of cells and is an active place (thereby requiring a lot of blood vessels to provide nutrients to the busy cells). At the very back of the retina are the photoreceptors, which are the cells that transduce light into electrical activity. The location of photoreceptors at the back isn't a problem because the stuff up front is transparent, so light easily reaches all the way to the rear. Embedded in each photoreceptor are millions of molecules called opsins or photopigments that absorb light energy. This kicks off a cascade of reactions that result in a change in the electrical potential across the cell's membrane. The change in voltage is a signal—it means the photoreceptor has detected light.

There are two main types of photoreceptor: rods and cones. (The names are based on the shape of the cell.) Rods have only one type of opsin and work better in dim light, so they provide night vision. Cones provide color vision. Cone opsins come in three different versions and are most sensitive to a specific range of frequencies. The colors you see are just different frequencies of light (or wavelengths, if you prefer—wavelength is inversely proportional to frequency). You can think of the three types of cones as red, green, and blue detectors, since their frequency range peaks in those colors; signals from the cones determine the frequency of light and hence its perceived color.

Cones are tightly packed into the fovea, the central part of the retina. This is where we have our highest resolution—we can make out the smallest objects when we look straight at them (or in other words, when the light from the object hits the fovea). Cones are scarce outside of this region, which means our peripheral vision—the part of the visual field surrounding what we're looking at—does not offer much in the way of color vision.[8]

Humans have about 125 million photoreceptors per eye. Since rods predominate outside of the small central region, they have to cover the largest amount of territory, and therefore we have many more rods than cones—120 million rods and 5 million cones.

Photoreceptors send their signals to other kinds of cells in the retina that perform various processing functions. For instance, signals from rods are combined, which increases the sensitivity of the periphery to light, thus improving night vision. Finally, retinal cells called ganglion cells carry the retina's output in the form of trains of action potentials to other parts of the brain. The axons of ganglions cells form a thick nerve called the optic nerve. This nerve is also known as the first cranial nerve, though most people just call it the optic nerve. There are two—one for each eye. Each optic nerve contains a little over a million axons, so the information coming from the roughly 125 million photoreceptors gets narrowed down and processed a bit before it even gets out of the retina. The information travels to a processing station in the thalamus called the lateral geniculate nucleus, and from there to the cortex. The first cortical region to receive visual information is called the primary visual cortex or V1 (also known as striate cortex from the white stripe of myelinated axons contained within it). From there the visual data splays out to various cortical regions. This is the neural basis of vision.[9]

It is already obvious that developing a visual prosthesis presents many more challenges than an auditory one. For one thing, you're dealing with millions of photoreceptors rather than just 25,000 hair cells. The dozen or so electrodes in a cochlear implant comprise a small percentage of the normal complement, but it doesn't take much to create a meaningful audio signal. A small number of frequencies or frequency bands won't give a listener a natural sound with all its rich nuances, but they're enough to provide a good deal of information. What can you get from just a few electrodes carrying light signals? As Penfield and others have discovered, point-like stimulations in the visual cortex produce flashes or stars of light. A few electrodes give you a few flashes. You could determine if its night or day or if someone shined a flashlight in your face, but not much else.

But every so often a flurry of reports surface in the news about bionic eyes, artificial retinas, and the warmly optimistic blind-can-see-again stories. I'm skeptical. Most of the feel-good stories are anecdotal. Researchers, like politicians, exhibit a certain amount of selectivity when it comes to their record—they spend most of their time talking about the one or two great successes.

Progress is definitely being made, however. What gives rise to hope of success is the brain's topography—the orderly arrangement of elements in the visual system. The retina forms a map of the visual world, and this map is maintained throughout the processing in the brain.

To explain what I mean by that, consider an example. Let's say you're looking at a cat. Your eye's cornea and lens form an image of the cat on the retina (helped, if necessary, by the additional refraction of eyeglasses or contact lens). Retinal cells that are stimulated by the image of the cat's whiskers are spatially close in the retina to cells that are stimulated by the mouth, which are close those of the nose, and so forth. Let's call the output of the retina at these points A (whiskers), B (mouth), and C (nose). The synaptic connections between the retina and the lateral geniculate nucleus maintains this orderly arrangement, so points A, B, and C in the lateral geniculate nucleus are also neighbors. The lateral geniculate nucleus projects its output to V1 with the same geometry, so again A, B, and C are close to each other. And the same is true for other points in the image. There is an orderly mapping of the visual elements, or, as neuroscientists like to say, of the receptive fields of neurons.[10] In other words, there is a map of the visual field in the neural processing stations.

The real story is actually a little more complicated than this because neurons in later stages grow increasingly specialized, and respond only to certain kinds of stimuli. It's not like there is a little person inside your head watching a screen somewhere—the map doesn't create an image for someone to view, it's just the brain's way of keeping track of what's what, and where it goes. Similar to the auditory system's tonotopic map, the visual map helps the brain with information processing tasks and perception.

What it means to prosthetic developers is that at various points in the system you can effectively insert an image—an electrical representation of one, that is. All you have to do is stimulate in the right places, in the right way.

* * *

A number of diseases result in degradation or sometimes complete loss of eyesight, but two of the most common are retinitis pigmentosa and macular degeneration. Retinitis pigmentosa actually refers to a class of genetic diseases which attack the retina. It gets its name from the dark deposits, or pigments, that are produced. In many forms of the disease, the first symptom is a diminution of peripheral and night vision, which you can guess is caused by a loss of rods. In other forms of the disease the cones are also affected. Either way, it is a progressive disease—it gets worse over time—and although total blindness is uncommon, the loss of vision is often severe. Macular degeneration is a disease that mostly affects seniors, and is the leading cause of blindness in the elderly. It is progressive, like retinitis pigmentosa, but it attacks the central region of the retina, known as the macula. This is where the fovea is located (the fovea is at the center of the macula), so macular degeneration results in a loss of central vision. Total blindness is rare but the loss of the high-resolution part of vision makes reading and similar tasks difficult or impossible.

When the retinal photoreceptors are causing the problem you can bypass them, just like cochlear implants that fill in for the absent hair cells. If the ganglion cells are intact you can stimulate the retina, and the optic nerve will carry the signals to the lateral geniculate nucleus and the visual system takes over. (The retina is actually part of the brain, at least developmentally—it arises from the same tissue in an embryo that gives rise to the brain—so retinal stimulation is brain stimulation, in a manner of speaking.)

You can also stimulate the optic nerve itself. Researchers have tried stimulating this cylindrical bundle of axons using a cuff electrode but it is difficult to access the interior fibers and decide which ones to stimulate.

The retina is the most accessible part of the visual system. Conceptually, a visual prosthesis is easy: use a device similar to a digital camera to convert an image into an array of electric currents that represent brightness or light values at certain points, then inject these currents with electrodes positioned accordingly in the retina. Sounds simple, in theory.

Actually, some of the best early work in visual prosthetics stimulated the cortex rather than the retina. William Dobelle, a researcher at the University of Utah and founder of the Dobelle Institute, wasn't the first in this field but his work is the earliest sustained, impressive effort to create artificial vision by electrical stimulation. Stimulating the cortex means doing a craniotomy, which is unfortunate, but a cortical prosthesis can work even if a disease or injury has ravaged the retina or optic nerve. From 1970 to 1972 Dobelle and his colleagues performed test stimulations on the visual cortex of volunteers who were undergoing neurosurgery for various other purposes (such as tumor removal). These patients weren't blind, but in the course of the operation their occipital lobe was exposed, and they willingly submitted to the tests. Dobelle explored parameters and positioning of electrodes. Then, in 1972, he began stimulating the visual cortex of blind volunteer patients. After developing an electrode array and camera system, Dobelle started implanting patients in 1974. The topography of the visual system makes it possible to stimulate the cortex with a reasonable hope of producing interpretable images.

Some of the implants were temporary, but others have lasted for decades. In 2000 Dobelle published a paper describing the results of one patient implanted in 1978.[11] This patient was 41 years old when implanted. He had lost vision in one eye due to an accident at age 22, and the other eye in another accident 14 years later. The electrode array consists of a rectangular arrangement of 64 electrodes on a platinum foil. The platinum electrodes are 1

millimeter (0.04 inches) in diameter. When an electrode is stimulated, the patient reports phosphenes—light flashes. The
phosphenes somewhat reproduce the orderly arrangement of the
array, with the phosphenes mapping to the equivalent of a few
square inches of space in the patient's left visual field at arm's
length. This is rather a small window of vision—think of a space
not much bigger than your hand with your arm fully extended—
but it is far better than nothing.

A miniature black and white movie camera provides the
images. It is mounted in the frame of a pair of eyeglasses. A
lightweight computer processes the data and uses signals carried
by wires to control the stimulation of the electrodes, trying to
adapt the predominant contours of the image to the array. At first
the patient couldn't identify symbols such as letters or numbers,
but with practice he became capable of reading certain symbols
when they're presented on 6-inch squares at 5 feet. Among other
accomplishments, the patient has also learned how to perform
some measure of navigation in rooms. Dobelle could adjust the
computer algorithms controlling the stimulator based on subjective reports of what the patient saw, fostering further improvement.

It's a start. Several avenues of improvement are possible,
including the use of more electrodes and more comprehensive
image processing software. Work has continued on cortical
stimulation, as described later in the chapter.

Many people would prefer to avoid a craniotomy. In the 1990s
another approach to visual prosthetics began to gain attention.
Researchers carefully examined patients with retinitis pigmentosa
and confirmed that they could invoke phosphenes by stimulating
the retina, even after the photoreceptors were long gone. This
means that a retinal implant—an artificial retina or "bionic eye"—
is feasible in these patients.

Any sight-impaired patient with a functional optic nerve is a
candidate for this kind of prosthesis. This includes most patients
with macular degeneration. Unfortunately, there are numerous
disorders in which this is not the case. Glaucoma and diabetes, for

example, commonly affect the optic nerve.

Recall that the output cells of the retina—the ganglion cells—lie in the front of the retina rather than at the back. If you place an electrode array in front of the retina you can easily stimulate the ganglion cells, but then the electrodes also protrude a bit into the eye. Going behind the retina—called a subretinal implant—eliminates this protrusion, but you don't have direct access to the ganglion cells.

It sounds uncomfortable either way, and requires surgery to get into the eye. But microsurgical procedures on the eye have become routine. Cataract surgery, for instance, requires removal of the lens of the eye. Often an artificial lens made of plastic is inserted in its place. Making incisions in the eye, though a delicate operation, presents no problem for a skilled ophthalmologist.

One of the most prominent researchers in the field of retinal implants is Mark Humayun, a professor at the University of Southern California and a member of the Doheny Eye Institute. Humayun and his colleagues have been laying the groundwork for retinal implants for years. Once researchers learned the ease with which phosphenes could be elicited from patients with degenerative eye diseases such as retinitis pigmentosa, the next step was to study what kind of subjective sensation a person would experience while being stimulated with multiple electrodes. Would an array of electrodes produce distinguishable spots of light that could form an image, or would it just produce an amorphous blob of light?

In a paper published in 1998 in the journal *Eye*, Humayun reported the results of testing various multiple electrode configurations in 14 volunteers, 12 with retinitis pigmentosa and 2 with macular degeneration. Under local anesthesia Humayun inserted the electrodes and positioned them on or slightly above the retinal surface. The smallest device consisted of two or three electrodes, and the largest array used was a 5 x 5 grid. The electrodes were connected by wire to a computer that controlled the stimulation parameters. All 14 patients reported phosphenes, and 13 of the 14 were able to localize the light to the part of their visual field corresponding to the site of retinal stimulation. Humayun also found that patients were able to resolve stimuli from different electrodes as long as they weren't too closely spaced. This means that retinal implants could potentially be used for object recognition, reading, and similar visual tasks.

As work progressed, some researchers became convinced that retinal or cortical implants could work. A permanent implant could perhaps provide a functional degree of long-term artificial vision.

* * *

Now is a good time to discuss biocompatibility. I have mostly avoided this issue up until now because the majority of electrodes have been temporary residents, or they didn't penetrate too deeply into tissue, as was the case for cochlear implants. But leaving little bits of wire in the brain or eye for decades raises important questions about long-term physiological effects, as well as the capability of chronically implanted electrodes to maintain their functionality.

In general, the body doesn't welcome intruders. Foreign objects near the surface get squeezed out eventually, and if an object such as a bullet is deeply embedded, the body tends to wall it off with scar tissue. And of course small intruders elicit attacks from the immune system, which causes inflammation and other problems.

The brain is sometimes called an immunoprivileged organ, which refers to a tolerance of foreign bodies. When scientists peer at tissues and fluids of the central nervous system they don't find very many white blood cells, which are the rank and file of the immune system. The reason for this is the blood-brain barrier, which restricts the movement of substances into and out of the brain. Obviously the blood has to provide nutrients, but the other stuff that travels in blood vessels is mostly blocked.

But that doesn't mean the brain is a quiet place, immuno-logically speaking. It can certainly mount an immune response to invaders such as viruses. Although white blood cells might not be constantly patrolling the brain, under certain conditions immune cells can and do enter. In addition, the brain has its own method of attack. A group of cells in the nervous system generically called glia perform a number of important functions, and certain members of this group, called microglia, surround and digest things such as dead tissue or intruders.

Despite relative immunoprivilege, an electrode in the brain (or eye) is not kindly received. Although immune responses and inflammation tend to be less vigorous, the nervous system is capable of encapsulating little slivers of metal if granted a long enough period of time to work at it. This process, sometimes called glial scarring, increases electrode resistance. In other words, it limits the flow of current from the electrode to the neurons. To compensate, the stimulation current can be increased, but too much current will produce too much heat, which causes tissue damage, as well as killing the cells by other means, such as initiating chemical reactions between the metal and surrounding tissue. Gradually the resistance becomes too great and the electrode loses its functionality.

Electrode manufacturers deal with this issue by using materials that are relatively inert and provoke the least amount of rejection. Metals such as platinum work well, and have generally been used in cochlear implants. You can also coat the electrode with various polymers that fight off encapsulation, but the electrode still has to conduct electricity even with these coatings or the whole thing is pointless.

There is also the issue of tissue damage. This has been discussed earlier in the sections that described the insertion of electrodes beneath the cerebral cortex, but even tiny electrodes implanted on the surface of the brain or retina can cause damage over the long term, especially if they provoke any kind of inflammation or chemical reaction.

Materials commonly used for electrodes these days include silicon and certain kinds of conductive ceramics in addition to metals such as platinum, gold, and silver. But all materials have disadvantages as well as advantages. No perfect substance has yet been found.

These issues have yet to be fully resolved. And the larger you make the electrode arrays, the worse the problems become.

\* \* \*

Researchers began to develop retinal implants soon after their promise became evident. But there were a lot of details to be ironed out. One of them is known as adaptation.

The visual system is mostly interested in change. This is true in general of all our sensory systems. Sensory systems adapt, and if nothing changes, there is nothing to report. This means the response to a steady stimulus gradually fades. Even an object clearly in the visual field will fail to be registered if it stays perfectly in one spot.

You might be suspicious of that last statement. If you hold your eyes steady and look at a motionless object, it doesn't disappear. But what you're not aware of is that your eyes are constantly making tiny movements. If they didn't, motionless objects would vanish. In 1952, two groups of researchers independently discovered this when they developed a way of presenting a constant stimulus to the eye.[12] (Think of a smudge on a contact lens—as the eye moves the smudge moves.) In a few seconds a truly constant stimulus disappears!

Which means you've got problems if you present a constant image as a continuous electrode current. Neurons respond to changes, not steady states. What you've got to do is stimulate at a certain frequency rather than all the time. What frequency do you use? You have to find the best one by trial and error.

Since the current spreads in tissues—which are bathed in salty, conductive solution—you have to worry about stimulating things you don't want to stimulate. This is as true in the retina as it is in the cortex. If too much current bleeds to other regions you'll be lighting up the whole visual field rather than producing a perceptible image. You'll also want to try to confine the stimulation to the ganglion cells near the electrode instead of stimulating axons from other ganglion cells. (Electrode stimulation is capable of eliciting action potentials in the cell bodies of neurons as well as their axons.) Those axons might be nearby but the ganglion cells they came from might be located far way; if you stimulate the axons in addition to nearby ganglion cells, the prosthetic user might see a flash of light around the center of the visual field and a simultaneous one somewhere far away.

Heat and current are also important considerations. Too much current and the device gets too hot, burning the surrounding tissue. This becomes a big problem when you incorporate a lot of electrodes in a small space, especially if you have to use additional current to compensate for glial scarring. Any associated electronics you implant can also create irritation and discomfort.

These are just a *sample* of problems to be addressed. And they are in addition to the general problems of visual prostheses: digitizing images and stimulating enough electrodes to create a perceptible visual experience.

Several groups of researchers are working on developing a visual prosthesis for the retina. Perhaps the best known of these is the Artificial Retina Project, which has involved a large number of researchers working in public and private laboratories. The list includes Humayun at the Doheny Eye Institute as well as researchers from the California Institute of Technology, North Carolina State University, University of California, Santa Cruz, and from five Department of Energy national laboratories— Argonne, Lawrence Livermore, Los Alamos, Oak Ridge, and Sandia. A company called Second Sight Medical Products, Inc., based in Sylmar, California, is sponsoring the clinical trials.

The project has developed three devices. The first, known as Argus I, has 16 electrodes.[13] Testing began in 2002 with six patients. Using this experience, the developers produced Argus II, a 60-electrode implant. Clinical trials began with this device in 2006.

The electrode array rests on the surface of the retina (an epiretinal implant). The stimulation area covers a square area of 5 millimeters (0.2 inches) per side. As with Dobelle's cortical implants, this is a small window constituting a few degrees of space (recall a circle has 360 degrees, so a few degrees is not a large span). And 60 electrodes don't provide detailed vision.

Images come from a miniature video camera mounted on eyeglasses. The camera's output is cabled to a computer that processes the data. The output of the computer is wired to the glasses where it is transmitted to an antenna attached to the implant. A tiny electronics package sends pulses of current through the electrodes as directed by the computer's instructions. A battery pack worn at the user's belt provides the electricity these components require.

The computer is an important part of the device. You've got to compress a ton of information about a scene into a paltry number of pixels. Which information do you pick out? Which electrodes do you stimulate? You need to decide how much of the scene you want to try to portray, and how to represent it in an optimal manner. Imagine you're looking at a picture of something—a face, a landscape, a living room—and your task is to create an identifiable miniature by picking out 60 points. But it's even harder than this: you've got geometrical constraints because you have to present the data topographically to the retina. But remember that the responses to adjacent electrodes might get smeared together. And the visual system in the cortex has its own way of doing things, and we still don't understand exactly how it works.

The small electrode array also poses a significant design challenge. Not only does the array have to be made of materials that can withstand a salty environment—and evoke as little inflammation as possible—it also has to fit onto the curved surface of the retina. The electrodes used by the Artificial Retina Project are housed in a thin film made of soft plastic, sort of like a miniature version of a contact lens (except it's worn *in* the eye).

It's amazing that it works at all. In some cases—particularly those described in press releases, newsletters, and articles written by the researchers and their funding agencies—the results are dramatic. A newsletter called *Artificial Retina News*, issued in 2009 by the U.S. Department of Energy, describes Kathy B., a participant in the Argus II clinical trials:

A highly functioning person with a job, family, and household to tend to, Kathy B. doesn't feel like she's blind despite the fact that she's had no vision for about 15 years. Nevertheless, she was excited when she recently was able to find the full moon in the dark, nighttime sky. "We were out walking, and I looked upwards, scanning my head back and forth," Kathy B. explains. "All of a sudden, I saw a big flash, and I asked my husband, 'Is that the moon?'" It was, and she began thinking about how long it had been since she'd last glimpsed that luminescent celestial body.

The clinical trials for Argus II went fairly well. Writing in MIT's *Technology Review*, journalist Duncan Graham-Rowe cites Robert Greenberg, one of the founders of Second Sight, as saying patients can find and identify some kinds of objects, including people, and follow their motion. Some patients can read if the print is large enough.[14]

With these results, Second Sight managed in 2011 to obtain approval to market the Argus II in Europe. The device costs the equivalent of about $100,000 in U.S. dollars. In a press release dated 2 November 2011, the company celebrated the first commercial implantation of their retinal prosthesis (as opposed to investigatory or research implantations). The operation took place in Pisa, Italy. The company has also submitted an application to the FDA for approval to sell the device in the United States. As of July 2012, the FDA has yet to make a decision. Researchers involved in the Artificial Retina Project are working to upgrade the electrode array to 200 or more electrodes.

Other projects for retinal prostheses include a subretinal implant developed by the German company Retina Implant AG. This company takes a wholly different approach—instead of using a camera to digitize visual information, they implant a photo-sensitive microchip under the retina and use the eye's own optics to form the image. The microchip is 3 mm (0.12 inches) in diameter and consists of 1,500 pixels. The pixels contain a photocell that converts light falling on it into electricity and a stimulating electrode. There is also the need for an external signal booster, since the current output of the photocells is insufficient to generate perceptible stimulation.

The company has begun clinical trials and made headlines in a lot of European newspapers in May of 2012 with success stories. It's an interesting approach—you don't have the problems associated with external cameras and image processing, but you're also limiting yourself to whatever information falls on the retinal implant. Since the implant is small it doesn't cover a lot of the visual field. However, people don't have high resolution vision except for the central region anyway.

A company called Optobionics developed a similar chip and began testing in the 2000s. But in 2007 the company went bankrupt. Although it may eventually get revived, it's a lesson and a warning that conducting a business on the frontier of medical research isn't easy.

Work on cortical visual prostheses has continued even though the pioneer, William Dobelle, died in 2004. One research group looking to develop cortical implants is the Neural Prosthesis Lab at the Illinois Institute of Technology. These researchers are testing a 16-electrode array that they hope to scale up to 600-650 electrodes, possibly more. Such an array in the cortex could provide vision over a considerable number of spatial degrees. But the cortex is wrinkled with a furrow (sulcus) here and there, and the hemispheric surface is curved. The geometry of the occipital lobe, where the visual cortex resides, is particularly complex. Installing an extensive array on the cortex may prove exceptionally difficult.

* * *

How many electrodes would it take to approximate "normal" vision? Surely you don't need every one of the 125 million photoreceptors. What's the minimum number of electrodes that can produce a level of artificial vision sufficient for most daily visual tasks?

The standard for visual acuity is generally measured in terms of a 20-foot distance (which is about the length of the old eye exam rooms). Most people with no visual problems have 20/20 vision—"normal" vision—which means that at a distance of 20 feet they can see what other people on average can see. The higher the second number, the greater the reduction in acuity. A person with 20/50 vision can see things at 20 feet that a person with normal vision can see at the much farther distance of 50 feet. Better than average acuity corresponds with a lower second number, so that a person with 20/15 vision—and there are some who have this acuity—sees at 20 feet what a person with normal vision has to be 5 feet closer to see.

A person with 20/20 vision can generally detect objects or textures that are as small as 1/60th of a degree in terms of visual angles.[15] This corresponds roughly to a distance of about 5 micrometers (0.0002 inch) on the surface of the retina.[16]

It's difficult to build electrode arrays this small and it's not at all clear that electrodes of this size could effectively stimulate a pinpoint region. Current spreads in the salty solution of the eye and brain, and there is no direct connection between the electrode and the cell or cells it is trying to stimulate, unlike in real neural networks where communication occurs across a small junction (the synapse).

Is it possible to develop a retinal implant with this level of precision? Probably not with today's technology, but don't bet against it happening in the future. Nanotechnology applications deal with objects many times smaller. Although nanotechnology is an overhyped field of research if there ever was one, people have been making progress with electronic miniaturization for a long time. The density of components on integrated circuits has ballooned from thousands in the 1970s to few billion now, approximately following Moore's law—named after Gordon Moore, one of the founders of Intel Corporation—which states that the capacity of chips doubles about every 18-24 months.

Suppose you could design and build a visual prosthesis with a potential resolution comparable to normal vision. Further suppose you've developed software to extract the relevant features of the signals from a video camera and you can stimulate the electrodes in a way that the brain can interpret. These aren't simple problems but they're surmountable. Would the user have "normal" vision?

Alas, the answer is no. There are additional complications.

Eye movement, for example, is a complication. About three times a second your eyes abruptly move as you transfer your gaze from point to point. Both eyes move in a coordinated fashion. These sudden movements are called saccades (from a French word meaning twitch).[17]

Here's the thing: you don't notice the movement. Roughly three times a second the world seems to fly past you but you don't sense this relative motion between you (your eyes, actually) and the world. It wouldn't seem to be any different than looking with a fixed gaze out the window of a moving car or train—except it is, or at least you perceive it much differently. When you look out the window of a moving vehicle the road and traffic signs and roadside trees seem to sweep past you, but when the same relative motion occurs during a saccade you remain unaware of it. Your visual world is stable despite these frequent eye movements.

But you're not closing your eyes during every saccade. Somehow the visual system has to compensate for the apparent motion that occurs during saccades. It's not hard for the brain to know when one is coming—the motor system is under the brain's control, so the visual system can get a warning from the part of the brain that moves the eyes. But neuroscientists aren't sure how the visual system adjusts so that the resulting apparent motion is ignored.

Mimicking this kind of mechanism in visual prostheses isn't going to be easy. In the relatively simple retinal and cortical implants constructed thus far, users simply keep their eyes as still as possible while they're looking at something. When they move their heads (or just the camera attached to the eyeglasses), the scenery changes and there's some delay as the software analyzes the new scene and the electrodes begin responding to the update, but it's not a big problem because the user is looking at an extremely low-resolution image anyway.

But what if the prosthesis delivered high-quality images?

Users couldn't move their sightless eyes, otherwise the motor system would send a warning to the visual system to ignore the apparent motion, but there wouldn't be any apparent motion because the camera didn't move. That would cause a great deal of confusion, and possibly a nauseating sensation akin to seasickness. Users would be forced to coordinate their eye movements with head or camera movements.

Implanting photodiodes under the retina, as performed by Retina Implant AG and the old Optobionics, would seem to obviate this problem. Because the image shifts with the eye, it should work as normal. And it would, if the visual system was working normally in these patients—but we know it isn't. In most cases the saccade signals and the resulting compensation are probably comprised.

This isn't the only problem that next-generation visual prostheses must solve. Features of normal vision include color, form, motion, and depth perception, yet researchers have so far focused their attention on one—form, or object recognition.

That's understandable because object recognition is perhaps the most important. But if you want "normal" vision you need the rest. Color relies on the signals of three different cones, and while it wouldn't be difficult for the software of a visual prosthesis to break the signal into its color spectrum, it *will* be hard to know which ganglion cells to stimulate for which color. Motion might not be difficult if you can learn how the retina usually goes about encoding it, and if you can solve the saccade problem. Depth perception requires two eyes—although you can get clues about depth from a single eye, the brain combines the inputs from our two slightly displaced eyes to get a stereoscopic view of the world. To reproduce that in artificial vision, you need two cameras or two sets of photodiodes.

Here's another problem, and it's a big one to those in which it applies. Certain functions don't develop properly unless the nervous system gets to practice for a while during specific times in the person's early development. These times are called critical periods, and if you're deprived during one of them, you may never be able to operate at a mature level. Language is apparently one such function (as mentioned earlier, this is one of the motivations for implanting cochlear prostheses in children as soon as possible), and vision is another. Researchers David Hubel and Torsten Wiesel showed many years ago that the visual system of kittens fails to organize properly when deprived of visual opportunities at crucial moments during development. The same seems to be true of humans; re-learning to see after a lengthy period of blindness is difficult but doable, but learning to see for an adult who has been blind since birth is probably not possible.

Yet there is so much hype for visual prostheses. Here's an example: "Researchers in the US predict that within five years they will be able to help some blind people to 'see'. After successfully transmitting simple electronic images directly to 15 blind people's retinas, they hope to make devices similar to the one used by blind engineer Jordi Laforge in *Star Trek: The Next Generation*." This is from a story in the 7 November 1998 issue of *New Scientist* that is referring to one of the early tests by Humayun's group.

In a way, the story's prediction proved true—some blind people have been helped to "see." But only in an extremely limited fashion. The tone of the story implied much greater things were on the horizon—five years from 1998 would be 2003—yet today no visual prosthesis can even modestly approach normal vision, much less anything out of science fiction.

Some critics even wonder if retinal implants do much good at all on their own. Part of the success could be due to healing of the retinal tissue, perhaps aided by the surgery or the stimulating current or some other poorly understood reason. This issue has been raised because some patients report improved vision in parts of their visual field corresponding to retinal areas somewhat distant from the implant! (Remember that many patients with retinitis pigmentosa or macular degeneration have some degree of residual vision even in late stages of the disease.) The greatest successes in visual prostheses might be due more to variability in the course of a disease such as retinitis pigmentosa—the progression isn't necessarily steady—as well as some kind of healing process induced by the implants or its electronics.

In my opinion, further research in visual prosthetics is worth pursuing. The theory behind these devices is sound, and those that have already been built, while limited in function, have managed in some cases to evoke images in the visual system of users, in addition to whatever healing properties are also taking place.

William Dobelle once described the future of his cortical implant as a series of phases: "Development of implanted medical devices such as this artificial vision system progresses in three stages. First there is speculation, then there is hope, and finally there is promise."[18] Yes, but will such devices ever live up to their promise?

I believe the answer to this question depends on expectations. There's little chance that any visual prosthesis will be able to restore normal vision to patients in the near future. This doesn't mean that these devices won't ever make dramatic improvements in the quality of people's lives—some of them do that now. Although visual prostheses haven't achieved the commercial success of cochlear implants, some users have regained simple visual functions they had thought they'd lost forever. It's all about managing expectations.

But every few years you find headlines in newspapers, magazines, and web sites that make claims such as, Implant Cures Blindness! Like acid reflux, these stories keep popping up long after they should have been history. Writers of these articles *should* know better—and editors and publishers *do* know better—but in the year or two between articles I guess they assume readers will have forgotten the previous miracles and be ready to accept a new batch. It's a shame because visual prostheses won't be able to live up to these extreme promises, and that could kill the whole field.

Other strategies to restore vision include gene therapy, which involves the replacement of faulty genes that cause inherited diseases such as retinitis pigmentosa, and the use of stem cells, in which new cells are grown to replace those that have been lost.

Nobody yet knows which if any of these treatments will be successful, but electrical stimulation of the retina and visual cortex is a strategy that will continue to be explored.

# 8
# Deep Brain Stimulation: Relieving the Symptoms of Disease

After the go-go era of the 1960s ended, the majority of researchers interested in electrical stimulation of the brain contented themselves with a more mundane but still important list of research topics: prosthetics (the subject of the previous chapter), stimulation of various nerves for prosthetic purposes as well as relief from some kinds of disorders, and diffuse stimulation of the entire brain by exposing the head to electric fields. Craniotomies were passé. The risk of infection, inflammation, hemorrhage, and brain damage caused by invasive electrodes didn't seem to be offset by any great advantages. Drugs had become the weapon of choice to fight psychiatric and neurological disorders. For those patients suffering from diseases not yet treatable with medication, it probably seemed worthwhile to wait until the right drug was discovered. Drugs conquered all, or would, eventually.

But there are several diseases that can become so debilitating they're terribly hard to live with. One is epilepsy. Although many patients are helped by one or another of the antiepileptic pharmacopoeia, and milder forms of the disorder don't create significant hardship, severe cases of epilepsy make life exceptionally difficult. Activities and mobility can be heavily restricted, and emotional or psychological disturbances can often arise from temporal lobe seizures. Even today, drugs fail to work in some cases, as they did for Brad, the patient I described in chapter 1. As we've seen throughout this book, epilepsy has driven a lot of research in the field of electrical stimulation of the brain.

Another problem that can be tough to deal with is chronic pain. Here again, pain is generally treatable with medications, and persistent pain usually has a source that can be addressed with an operation or some other medical intervention. But what if it isn't?

What if the pain continues, unresponsive to painkillers, in the absence of any clear medical cause?

Perhaps in some cases the pain is psychosomatic—in the mind. This is sometimes viewed derisively: "It's not real, it's all in your head." I have a different point of view. If you *feel* it, it's real, whatever its source.

Neglecting the cases in which the patient is for some reason lying, persistent pain of a mysterious origin can have at least one possible source—the brain. You feel pain when certain nerves in the peripheral nervous system send signals to the brain. Recall that there are no pain sensors in the brain itself, but there are of course such sensors in the body. These sensors are often associated with free nerve endings, which are basically the bare end of an axon. (As opposed to more elaborate sensors of the body, such as the receptors that respond to touch which are composed of complex structures.) Tissue damage or other extreme conditions cause pain sensors to send signals through the spinal cord and into the brain. Pain is a form of tough love; it carries out an important behavioral function, warning us not to exceed our limits.

Except when the signals are false, or are warning you about something that can't be helped, in which case you feel pain but there's not anything you can do about it. The pain system of the brain is stuck in DO SOMETHING mode but you're helpless. This can happen in several ways. Sometimes pain signals arise due to injuries to the nerves or the brain, such as strokes or trauma, which cause a perception of pain in the body even though that part of the body may be perfectly fine.[1] In this case the pain is truly "all in the head" though it feels like it is coming from the body. (Pain arising from nervous system malfunction is sometimes generically called neuropathic pain.) In other cases, chronic pain may arise from an injury in the body but it's unfortunately part of an irreversible medical condition, such as certain types of cancer or some forms of bone or joint diseases.

Penfield rarely elicited painful experiences while stimulating the cortex of his patients. His data suggested that perception of pain in general may rely on structures deeper in the brain. Stimulating structures buried below the cortex is called deep brain stimulation, often abbreviated DBS. (But as we'll see below, DBS these days isn't necessarily subcortical.) Heath, Delgado, and their colleagues did a lot of it, though they didn't tend to call it DBS.

In the 1970s and early 1980s, electrical stimulation of the brain of human patients was mostly limited to last-ditch efforts to alleviate chronic pain.[2] Two primary targets were identified early on. One was a nucleus of the thalamus that processes somatosensory information (ventralis caudalis), and often referred to as sensory thalamus. The other was a ring of cells—"gray matter"—around a fluid-filled space called a ventricle[3], so this nucleus is known as the periventricular gray or a related site which is nearby (actually continuous with it) called the periacqueductal gray. The theory behind stimulating these sites was, and still is, vague. Stimulation of the sensory thalamus causes tingling and perhaps scrambles the pain signals. Research on rats in the 1960s showed that analgesia could be induced by brain stimulation in the periacqueductal region.

The periacqueductal gray we have met with before—Delgado elicited aggression in monkeys during stimulation at some currents and frequency parameters in this area. Other parameters—or stimulation at slightly different parts of the region—can produce the opposite result, particularly if they are inhibitory. (This sounds confusing, and I'm afraid it's going to get worse—nobody understands how and why DBS works. Additional confusion will be presented later in perplexing detail.) Stimulation of the periventricular or periacqueductal region under some conditions may also activate the body's opioid system—the same system that opium and morphine affects. It may also influence brain processes involved with processing emotion, taking the bite out of any pain the patient may be experiencing.

These procedures to treat chronic pain were mostly implant operations similar to those performed by Heath and Delgado—the stimulating electrodes were generally intended to stay put—rather than temporary stimulation during brain surgery, such as what Penfield did. This early DBS work and the treatments that have followed owe a lot in spirit as well as in practice to Heath and Delgado, though it has become unfashionable to cite the work of these controversial researchers.

A review of the early DBS efforts to relieve chronic pain was published in 2005.[4] This review examined the results of six studies: one published in 1977, four in the 1980s, and one in 1997. The six studies treated a combined number of 424 patients suffering from a wide variety of chronic pain ailments, though more than a third of patients had some kind of problem with their back. In most cases the site of stimulation was the sensory thalamus or the periventricular/periacqueductal region or both.

The surgical techniques varied, as did the technology over this 20-year period, but in most of the procedures the physicians implanted electrodes using stereotaxic instruments and tested their efficacy for some period before the stimulation system was completely installed. The stimulator was generally connected to a controller by a wireless transmitter/receiver.

The stimulus parameters—frequency, voltage, and so forth—varied from study to study. The reason, quite simply, is that these procedures were experimental and nobody knew what the optimal settings might be, or how the best choices might vary from patient to patient. That might make it sound as if humans were used as guinea pigs here. Actually, in a way they were, but these studies differed significantly from, say, the experiments of Roberts Bartholow on Mary Rafferty. Although the researchers in these six studies had to experiment with the parameters, they had some degree of confidence that the procedures would help at least a number of the patients. Thanks to the work of Heath, Delgado, and many others, there was enough data to show that brain stimulation can have important benefits.

You might be wondering about the use of experimental animals. Couldn't these experiments have been performed on actual guinea pigs first?

The answer is yes—and no. A lot of experimental work in electrical stimulation has been done in animals, some of which has been described in the previous chapters of this book. These data guided Heath and Delgado as well as the researchers who followed them. But eventually you reach a point where tests with animals aren't sufficient. This is especially true in fields of research involving perception or some other higher brain function. You can do experiments on animals that involve painful stimuli and learn something useful that may also apply to humans, leaving aside the hot-button issue of inflicting pain on sentient creatures, which nobody enjoys doing. But a detailed understanding of various sources of chronic pain and the way the human mind responds to it will not come from animal studies. Even among humans the individual variability is so great that generalities are sometimes difficult to find.

The success rate in these six studies was generally high. What constituted a "success" varied along with the stimulus parameters and other details of the procedures, but a high percentage of patients experienced some amount of pain relief. And when you're hurting *all the time*, you're probably going to find just about anything that relieves the pain to be worthwhile.

Successes could be broken into two categories: initial relief and chronic relief. The outcomes can unfortunately be quite different. Because bodies and brains adapt, the results you get from brain stimulation when you first turn on the current and the results you get some weeks or months later won't often be exactly the same. The success rate for initial stimulation was 305 out of 424—72 percent. Long-term follow-ups lasted anywhere from a few months to as long as 15 years in some patients. Unfortunately the success rate dropped considerably over time as the brain adjusted to its new environment. The number of long-term successes was 232, for a success rate of about 55 percent.

Cortical stimulation has also been employed for relief of chronic pain. That might sound strange in light of Penfield's findings, but Penfield's stimulation parameters and conditions are not the same as those for patients with long-term implants. When researchers began contemplating the use of cortical stimulation for the relief of chronic pain, their first idea was to stimulate the somatosensory cortex. Some researchers suspected that stimulating this region of the cortex would scramble pain signals in the same way that stimulating the sensory thalamus does.

But like many other perfectly reasonable and logical hypotheses in neuroscience, it proved wrong. However, in the early 1990s researchers scored a success when they shifted the electrodes to a nearby cortical region—the motor cortex. The pain relief was achieved with a low level of current, which is fortunate because otherwise the stimulation would have evoked unwanted movements. Why it works is even more mysterious than stimulation of the sensory thalamus and periventricular/periacqueductal gray, but it has become another tool in the pain-relief arsenal. Because the site of stimulation is cortical rather than deep it seems strange to classify the procedure as DBS, although some people call it that. This is also the case for another cortical procedure described later in the chapter.

Brain stimulation for relief of chronic pain is rare these days. It's still done in Europe, but not so much in the United States, and the FDA has not yet given its approval, which means all such procedures in the U.S. are done as experimental cases.

Much more common than DBS for chronic pain relief is spinal cord stimulation. It's the same idea—use electrodes to change how neurons process information. Most versions of these spinal cord stimulators consist of electrodes in the spinal cord plus a small pulse generator implanted in the back. A stimulator to control the process and a battery to power it can also be implanted, often in the abdomen, but some versions use a radio-controlled receiver that gets its instructions and power transmitted to it. Users adjust the stimulus parameters to suit their individual preferences, but commonly activate the stimulator several times a day for an hour or so. The first spinal cord stimulator got approval from the FDA in 1981, and several others have been approved since then.

This is an invasive procedure, with all the risks inherent in such operations, and the spinal cord is almost as precious as the brain, so it's not something to be taken lightly. A few percent of patients suffer complications—the exact number depends on your definition of "complication"—and catastrophes do occur during implantation, but these are fortunately rare. But if you've experiencing constant pain that you just can't shake and nothing else has worked, the risk might well be worth taking. The success rate of spinal cord stimulation for pain relief seems to be comparable to DBS, so most people are opting these days for the spinal cord rather than the brain.

Without any other applications, DBS might have gone the way of the dodo and nonpartisan politics. But in 1987, researchers found a real winner.

* * *

I met Ben in 1996 at a Veterans Administration hospital. He was a robust man with a thick and unruly thatch of gray hair covering his head—a Korean War veteran who looked distinctly uncomfortable in a place full of sick people. You got the impression he hadn't been sick a day in his life. Until, that is, a few years ago, when he began to notice a tremor in his left hand. Physicians had subsequently diagnosed him with Parkinson's disease (PD).

People have probably been suffering from PD throughout history, but the disease got its present name in 1817 when British physician James Parkinson published a report describing in great detail a kind of "shaking palsy." It wasn't a prevalent disease in earlier times, perhaps because it generally strikes people beyond the age of 60, and so the prevalence of PD has increased along with the average life span (though people much younger than 60 can also come down with the disease).

Most doctors characterize PD as a movement disorder. Symptoms vary among patients, but people suffering from PD have at least one of the following primary symptoms: resting tremor in a hand or foot that usually progresses to the other side (the tremor occurs when the affected limb is still or resting), bradykinesia (slow movement), rigidity, and postural instability or loss of balance. There are numerous secondary symptoms, and patients in later stages of the disease can experience cognitive impairment, delusions, and a host of other problems. In general it isn't considered a fatal illness, but while there are treatments to alleviate the symptoms, there is no cure.

Ben's hand shook noticeably when it rested on his lap. But it wasn't often resting—Ben used his hands a lot when he talked, and when he wasn't talking he often twisted his watch, which he wore on his right wrist, or he would form a steeple with both hands and tap his fingertips together, a little like one of those old television show detectives when thinking over a particularly difficult case. I got the feeling Ben was constantly in search of things for his left hand to do. When the hand was in motion the tremor faded.

His speech wasn't slow or soft as I had expected for a PD patient. Instead he spoke rather quickly, though he did stutter on occasion.

At the time when I spoke to him, deep brain stimulation for movement disorders was still in its experimental stages, although it was getting some attention in the news (in that seemingly ancient era before the Internet had become huge, the "news" meant newspapers, magazines, and journals). I asked Ben if he had heard about it. He nodded solemnly and then quickly changed the subject.

Like many PD patients, Ben was taking L-dopa, short for levodopa, a molecule that is a precursor to the neurotransmitter dopamine. What this means is that neurons can use L-dopa to synthesize dopamine, which a small but important subset of neurons use as a transmitter to signal other neurons. In other words, dopamine is the chemical that is released by the signaling neuron, crosses the synaptic junction, and attaches to specific dopamine receptor proteins in the target neuron's membrane. Consuming L-dopa elevates the brain's ability to make dopamine, which is lost in the course of PD. (Why not eliminate the middle man and prescribe dopamine instead of a precursor? The reason is that L-dopa can slip through the blood-brain barrier but dopamine can't.)

Researchers have discovered what causes the symptoms of PD: over the course of the disease, neurons in a subcortical nucleus called the substantia nigra begin to die.[5] These neurons are dopaminergic, meaning they use dopamine as a neurotransmitter. Apparently L-dopa alleviates the symptoms by boosting the output of the remaining neurons.

The big question in the search for a cure is what causes the neurons in the substantia nigra to die off. As yet nobody knows, though there is no lack of theories. Popular theories these days involve genetics or various environmental toxins, or some combination thereof. Perhaps PD is a family of diseases that have different causes, which may be one of the main reasons why researchers are having a devil of a time finding the answer. L-dopa is a temporary reprieve rather than a cure, and the drug gradually loses its effectiveness as more neurons die and there are fewer neurons left to carry the load.

A lot of PD patients experience "on" periods, in which they enjoy relief from the symptoms, and "off" periods when they don't. Ben told me he had been lucky because he hasn't had much fluctuation, but I suspected the roller coaster was starting and he was currently experiencing a lot more "off" periods lately— presumably why he was visiting the hospital, although I didn't ask.

Ever since researchers discovered that the loss of substantia nigra neurons is associated with PD, they have tried to figure out how this nucleus is involved in the motor system and what it does. A better understanding of the substantia nigra and its function could aid the search for improved treatments. Unfortunately—and you probably saw this coming—it's complicated.

The substantia nigra belongs to a group of interconnected nuclei known as the basal ganglia. In addition to the substantia nigra, the basal ganglia include the putamen, caudate nucleus (one of Delgado's favorite targets), globus pallidus, and sub-thalamic nucleus.[6] I'm not going to describe the circuit or how these nuclei are interconnected. I get the uneasy feeling that already in several parts of the book your eyes may have begun to glaze over, and if so, you'd definitely lose focus during any sort of in-depth explanation of basal ganglia circuitry. Nobody completely understands how it works anyway.

It's obvious the basal ganglia are involved in voluntary movement—when something goes wrong with this part of the brain, movement is affected, among other things. Voluntary movement occurs whenever you decide to take some sort of action. It would seem to be a simple thing to understand. Remember earlier when I likened the brain to a processor? Sensory data comes in, the brain makes a computation, and then the output—behavior—occurs.

Well, it's not so simple when you get into the details. Even if you've got a goal in mind—a decision that's probably made somewhere in the "higher" processing centers of the cortex—it's not easy to know how to best put the plan into effect. That's probably part of the basal ganglia's job. The body's muscles have to be orchestrated or you won't get to where you need to go, and the cortex leaves the details to the basal ganglia. Synaptically, one member of the basal ganglia (the caudate) receives information from the cortex, and this data gets processed throughout the basal ganglia and then gets outputted to the thalamus, which sends it right back to certain areas of the cortex. It forms a loop, in other words. The exact function of this loop is unknown but it seems to play a role in selecting which movement to make and/or co-ordinating the muscles to carry it out.

Since disruptions of this loop affect the motor system, re-searchers began trying to electrically stimulate parts of the loop to see if they could disrupt the disruptions. This is the purpose of deep brain stimulation.

I asked Ben if he'd consider DBS treatment. His face, which seemed to lack expression—a symptom of PD—briefly darkened.

"No," he said.

But then he became silent for a moment. I didn't interrupt him.

"I guess I might," he finally added. "If this thing really gets bad."

* * *

Parkinson's disease is one of the most common movement disorders. Another two movement disorders that unfortunately strike a lot of people are called essential tremor and dystonia.

Essential tremor, as the name suggests, is characterized by shaking or an uncontrolled rhythmic oscillation in part of the body. Tremor is generally a symptom of PD but essential tremor is a distinct disorder, and the shaking is usually different—whereas PD tends to involve resting tremor, the shaking in essential tremor tends to accompany voluntary movement. Essential tremor often affects the hands, but can also affect the head or other parts of the body. There are several different forms of this disorder and the etiology—the cause—is not well understood, and nobody can point to any specific part of the brain as the culprit in all cases.

Dystonia causes involuntary muscle contractions. These contractions are like spasms or cramps but are severe and tend to produce slow repetitive movements or twist the person into an abnormal posture. There are several different types of dystonia; the general cause is not known, though some cases appear to involve the basal ganglia.

These disorders are chronic but not fatal. If severe, though, they can seriously degrade the quality of a person's life.

Although essential tremor and dystonia are distinct from PD they are all movement disorders. They also have something else in common: all can be treated with DBS.

The story begins in 1987 when French surgeon Alim-Louis Benabid and his colleagues reported the reduction of tremor in movement disorders by stimulating a certain nucleus (ventral intermediate nucleus) of the thalamus. Benabid wasn't the first to apply DBS as an experimental treatment for Parkinson's disease and other movement disorders—beginning in the 1970s several researchers began testing stimulation of various brain structures such as parts of the basal ganglia, thalamus, and cerebellum—but Benabid and his colleagues began systematic testing that ultimate led to acceptance of DBS as an effective tool.

The idea of applying electrical stimulation to treat movement disorders was a natural progression from the use of electrodes in "lesion" treatments. Benefits can be attained by destroying a small amount of brain tissue in patients suffering from certain movement disorders, similar to treatments for severe epilepsy. If you eliminate the cause or part of the cause of the problem, the symptoms will improve even though you've lost the normal function of the destroyed tissue. To do a careful job of it and lesion the smallest area that will have the most beneficial effect, physicians in the 1960s implanted electrodes in patients and applied low-current stimulation to certain regions, testing the resulting reduction in symptoms. When they found the part of the brain that seemed to be bad, they let the current rip, zapping the tissue and killing the neurons.

But what if you kept stimulating instead of committing an irreversible destruction? Researchers discovered that in some cases high-frequency stimulation—stimulation at a frequency above 100 hertz or so—did the same job as a lesion. With DBS you didn't have to remove or burn part of the brain. (Which reminds me of Delgado's argument—see chapter 6.)

It means leaving the electrodes implanted in the brain, which as described in chapter 7 can be troublesome. It also means you have to provide some source of current and a mechanism to control its flow.

Today neurosurgeons have excellent tools to place electrodes precisely in the brain. Imaging devices such as magnetic resonance imaging (MRI) or computerized tomography (CT) help surgeons place electrodes in the desired spot. Gone are the days of Horsley and Cushing, when surgeons used cranial landmarks and an atlas, and hoped for the best.

A DBS system consists of an electrode, a wire, and a stimulator. The electrode is a thin cylindrical conductor, insulated everywhere except for the contact points near the tip at which current flows. Sometimes there are two electrodes, one for each hemisphere. The electrodes have four contacts which are 1.5 mm (0.06 inch) in length along the wire and have a diameter of 1.27 mm (0.05 inch). (Having four contacts instead of just one provides a greater flexibility in the stimulus parameters—it gives you more options.) The wire connects the electrode with the stimulator, and is generally placed under the skin. A stimulator about the size of a stopwatch contains batteries for power and controls the stimulus parameters. It is also implanted under the skin, usually near the collarbone but can also be placed somewhere in the chest or abdomen.

The U.S. Food and Drug Administration approved the use of a DBS device in the treatment of essential tremor in 1997, Parkinson's disease in 2002, and dystonia in 2003. The only company whose devices have been approved is Medtronic, a large medical technology company based in Minneapolis, Minnesota. (The FDA must approve each specific device before a manufacturer can market it.) Medtronic has been around for a while, and became well known for developing cardiac pacemakers. But Medtronic's DBS monopoly may soon come to an end, as St. Jude Medical of St. Paul, Minnesota is testing its own DBS device.

There are millions of patients worldwide who have been diagnosed with a movement disorder, but only a few percent—tens of thousands—have opted for DBS. Mild cases don't require anything drastic, and the risks of invasive surgery deter others. And some, such as Ben, may just not like the idea of electrodes in their brain, even though they are willing to swallow drugs that meddle with their neurons and sometimes have pretty awful side effects. Pills, yes; electrodes, no. (I lost touch with Ben so I don't know if he ever changed his mind.)

The risks associated with DBS are the usual ones for the brain stimulation procedures described earlier: bleeding, infection, and seizures. If bleeding is severe it can lead to a stroke. Other problems arise if the electrodes break or move, or if the subcutaneous wire and stimulator cause trouble. The percentage of patients who experience problems varies depending on the procedure, but it is generally in the 5-10 percent range.

Long-term stimulation of subcortical structures may also have serious drawbacks that are only beginning to be noticed. For example, some patients have reported a loss of memory. But it's difficult to attribute symptoms to specific causes; it could be DBS, or the disease itself, or a side effect of one of the drugs the patient may still be taking, or just part of the natural aging process. Still, long-term side effects of DBS bear watching closely, and they will probably be getting increasing attention over the coming years.

But for many patients DBS alleviates the symptoms of their movement disorder. It's not a cure, but at least it's reversible, unlike treatments that remove or destroy part of the brain. Drugs such as L-dopa for PD can keep the symptoms under control in the early stages of the disease, but their efficacy usually drops over time. The "off" phases increase and become less manageable. Patients elevate their doses in an effort to ward off this deterioration, but this also causes the side effects to expand in number and severity. L-dopa, for instance, increases dopamine production, which has extremely powerful effects in the brain. Dopamine neurotransmission plays a role in a variety of systems, particularly ones involved in decision making and reinforcement. An excess of dopamine can result in serious psychiatric problems. With DBS—when it works—patients can greatly reduce their medication, though they generally can't eliminate drugs entirely.

The site of stimulation depends on the particular movement disorder and on the patient's response to DBS, which of course varies. Patients diagnosed with essential tremor who opt for DBS will usually have the thalamus stimulated, whereas the target for patients with dystonia tends to be the globus pallidus, though sometimes it is the thalamus. In PD, common DBS sites are a particular part of the globus pallidus (the internal segment, often abbreviated GPi, which is a part of the nucleus that is slightly deeper and more toward the midline) and the subthalamic nucleus.

There's a big question associated with DBS that researchers have been scrambling to answer ever since the beginning: What is DBS actually doing to alleviate the symptoms? If we could solve this puzzle we could probably enhance the effects and optimize the procedure for each disorder and each patient, as well as perhaps extend the applicability of DBS. The problem is that no one knows for sure, although lots of people have opinions.[7] A great way to start a heated debate in a room full of neuroscientists is to ask how DBS really works.

One of the major obstacles in this research is that we don't know how the basal ganglia and the rest of the motor system work under normal circumstances, which makes it that much harder to deduce what's going on when something goes wrong, as with PD. When you add yet another complication—the injection of current from an electrode that is completely foreign to the system—it's no wonder the results are enigmatic.

The original hypothesis of how DBS works is conceptually simple, and I alluded to it earlier. Because high-frequency stimulation in some cases mimics ablation—the removal or destruction—of the stimulated site, it's easy to conclude that DBS inhibits the target. In other words, it keeps most of the neurons from firing many action potentials.

Sounds strange. Stimulating the brain means applying a voltage and causing a current to flow, which tends to excite cells—an effect neuroscientists call depolarization[8]—and leading to more rather than less action potentials. How could stimulation *inhibit* a target?

There are several ways this could happen. One way is called depolarization block, which involves a particular ion channel. This ion channel opens briefly to start the action potential process, but it is inactivated shortly thereafter, and the membrane must return to normal before it can open again. This mechanism prevents the neuron from firing too many action potentials or getting too much current. But it also means that if the neuron is constantly depolarized—if it is always excited—then these channels refuse to work and therefore no more action potentials are forthcoming. The neuron is blocked with depolarization.

Another way inhibition can arise is if the stimulation excites action potentials in certain neurons that are inhibitory, which are the neurons that usually squirt the neurotransmitter called gamma-aminobutyric acid (GABA) across synapses. GABA tends to reduce the membrane potential[9] of the recipient neurons, which causes them to fire fewer action potentials. If this is the case, then DBS may invoke a small amount of brain activity but mostly in neurons that suppress the others, resulting in a net inhibitory effect.

You can imagine all sorts of combinations of excitation and inhibition. Stimulated areas may contain both neurons and passing axons (and remember that action potentials can be evoked in the cell bodies of neurons as well as their axons). If the axons are stimulated they will take the excitation (in the form of evoked action potentials) out of the area, wherever their destination might be. Recall this was also a possible effect of retinal stimulation, and in general for most kinds of stimulation.

The situation is even more complex if you consider the downstream effects—no nucleus or region of the brain exists in isolation, for what would its purpose be in such a case? All parts of the brain are synaptically connected to *something*, which means if you stimulate a bunch of neurons then they'll deliver the message, whatever it may be, to other neurons, and so forth down the line.

So, what is DBS really doing? Researchers have conducted a lot of studies; they've analyzed the effects of stimulation on neurons growing in Petri dishes, experimented with animal models[10], and used sophisticated imaging techniques to look at the human brain in action. Much has been learned from these studies, but the bottom line is that there is no single hypothesis that explains the effects of DBS in every case. It seems to be a mixture of excitation and inhibition. In some cases it turns off overactive neurons, and in other cases it brings the activity more in line with normal rates.[11] It depends on stimulus parameters such as frequency and intensity as well as the exact site of stimulation.

The effects of DBS are probably not limited to excitation or inhibition. Neurons aren't just electrical components in a circuit, they are living cells with all the standard biochemical activities associated with life—complex reactions to make energy and perform various maintenance functions, and the expression of genes and synthesis of proteins as needed. DBS almost assuredly has an impact on a neuron's biochemistry, particularly over the long term as the cells adjust and adapt to their new circumstances, which cells always do. As yet this hasn't been widely studied but it needs to be. Some of the most important effects of DBS may be chemical rather than electrical.

* * *

Physicians treating movement disorders aren't the only people who've gotten interested in DBS. Psychiatrists have also started looking into it—but with a great deal of caution. Brain surgery and psychiatry have an unsavory past, at least in terms of our modern, politically correct world. The history of surgical interventions to treat mentally ill patients includes the liberal use of frontal lobotomies, pioneered by Egas Moniz—the guy who shared the 1949 Nobel Prize in physiology/medicine with Walter Hess. This liberal use, in modern terms, has grown to mean abuse, as psychiatrists began lobotomizing patients left and right to calm them—and seemingly taking away their personality and independence at the same time.

But psychiatrists who have heard about DBS couldn't help having their curiosity piqued. Psychiatrists, like physicians who treat movement disorders, must deal with patients who suffer from chronic and severely debilitating illnesses, and one odd finding about DBS in particular stood out: physicians using DBS to treat movement disorders noticed that the electrical stimulation can sometimes alter the patient's mood or emotional state. A trickle of case studies came out during the early years of DBS and managed to get some attention. Psychiatrists began saying, "Hmmm, I wonder...."

An early candidate for DBS treatment was obsessive-compulsive disorder, often abbreviated OCD. There are several reasons that people focused on OCD. For one thing, severe cases of OCD that failed to respond to all other therapies were being treated with surgical ablation—and researchers who studied DBS were getting the idea that high-frequency stimulation could in many circumstances mimic lesions. Another reason people targeted OCD is that the disorder appears to be associated with dysfunction in the same sort of cortex-basal ganglia-thalamus loops as movement disorders. The specific sites are different and the two kinds of disorders are clearly distinct, but people suspected the underlying mechanisms could be of a similar nature. If so, DBS might work for OCD as well as movement disorders.

One of the physicians who developed the application of deep brain stimulation to treat intractable cases of OCD was Benjamin Greenberg at Brown University. But he was wary. In a 2003 paper, he and his colleague Ali Rezai discussed the pros and cons, and cited as a precautionary note the old uses and abuses of frontal lobotomies. "Patients underwent that surgery without adequate long-term safety data and without careful characterization of their primary disorder. Tragic consequences were frequent. These remain a vivid reminder of the need for caution in this area."[12]

But severe cases of obsessive-compulsive disorder, like Parkinson's disease, can make a patient's life miserable. Experimental trials proceeded.

When many people hear about OCD, they might think about a germophobe who constantly washes his hands. But the disorder is much more than that. From a psychiatric standpoint, obsessions are recurrent thoughts or ideas that create anxiety, and compulsions are repetitive behaviors that a person can't stop or control.

Everybody has probably known someone with a quirky compulsion or two. When I was in the Air Force I knew a guy who refused to fly on an airplane without carrying a silver dollar his grandfather had given him. One time just before a flight he discovered to his horror that he'd left his lucky coin at the hangar, so he ordered a skeptical airman—me—to retrieve it at the last minute. As I was dodging taxiing airplanes on my way to the hangar I had an urge to find another silver dollar that I could give him, and then when he got back from the flight safe and sound I would produce the real silver dollar and say, "See? It didn't make a difference." But then I thought, *what if something happened....*

To be diagnosed as a psychiatric disorder, obsession-compulsion must dominate a person's life. It becomes almost ritualistic, and the afflicted person will spend hours counting something or checking something or washing something—and then spend the whole night worrying that he hadn't done it just right. *That* is OCD, not some silly superstition about a silver dollar. (Drug addiction—a compulsion to take drugs—falls under a different psychiatric category.)

A number of medications exist to treat OCD, but when these fail to work—which happens distressingly often—the patient may opt for more drastic therapy. Sites that have been stimulated for OCD include a tract of white matter (axons) called the internal capsule (which is located in the basal ganglia area), the caudate nucleus, and the subthalamic nucleus. The internal capsule site in these studies was in the ventral region (meaning the lower part, as opposed to dorsal, which refers to the upper) and quite near the ventral striatum, which is a region below the basal ganglia and often considered a part of the group. So this site is generally called ventral capsule/ventral striatum (vc/vs). These targets—vc/vs, caudate, and subthalamic nucleus—were chosen based on what has worked in ablation treatments as well as experiments and imaging that have attempted to trace connections and circuits that malfunction in OCD. The procedure is similar to that used to treat movement disorders.

The trials have gone fairly well, but there haven't been nearly as many patients as there have been with the movement disorders. Although OCD is unfortunately a common disorder, striking millions of people worldwide, the number of patients in such extreme condition as to warrant DBS is small. In keeping with this, the U.S. FDA gave special approval in 2009 for Medtronic's DBS device in OCD. It's called a humanitarian exemption—the requirement for loads of data and expensive trials was waived. The FDA offers such exemptions in situations where the device appears safe and effective but the manufacturer would be unable to recoup the expense of proving it (and thereby gaining standard approval) because there are so few potential customers. The limit is 4,000 patients per year. No more than 4,000 patients can use Medtronic's device for treatment of OCD under the FDA's humanitarian exemption.

Thus far the number isn't even close to that. A high percentage of OCD patients who have received brain stimulation have been helped, but it is important to point out that it's not a cure. As with DBS for movement disorders, the symptoms become more manageable and the medications work better. If DBS is stopped, the improvement fades.

Greenberg said it best in the following quote, cited in a Brown University press release dated 16 February 2011: "What DBS really does is make you into an average OCD patient." But an "average" OCD patient can often function in society and enjoy life despite the symptoms, whereas people who seek DBS are generally in dire circumstances. What makes successful OCD treatments so important is that unlike many other severe mental disorders, OCD can hit children as hard as adults. And once it's got you it doesn't tend to let go. Adults with OCD have likely had it for a while and will have it far in the future. It's no wonder that some of them are desperate.

* * *

Neuromodulation is starting to become a much more popular catchphrase than a mundane term like brain stimulation. It certainly sounds better: "We're going to modulate your neurons" is more appealing than "We're going to drill a hole in your head and insert electrodes into your brain". But implanting electrodes into the brain is what DBS is all about, and to many healthcare providers its successes have proved its worth and justified the risk in a significant number of cases. Now physicians and psychiatrists are casting a yearning gaze toward other prospective applications.

You might think that researchers' lack of knowledge of DBS's physiological mechanisms is a problem. It would certainly be beneficial to understand how DBS works because then it could be applied optimally and people would know, or think they know, which disorders it could help and which it couldn't. Not knowing this, almost any disorder is fair game. But that can actually be advantageous in a certain perverse sort of way. Without theoretical restrictions the sky's the limit and bold guesses might be rewarded. Or not. My point is that sometimes people are sure they understand something when they really don't, which leads to mistakes and lost opportunities.

One of the most widespread psychiatric disorders is depression. I've seen estimates that as many as 10-20 percent of the world's population will have to deal with at least one major episode of depression in their lifetime. This isn't just a case of feeling bad for a while or a little out of sorts. To be diagnosed clinically, it has to reach the stage where it seriously affects your quality of life and your ability to function in society. Some people experience depression but manage to recover on their own, or perhaps with the help of friends or a good therapist. Other people—probably the majority these days—take drugs such as Prozac, Zoloft, and others. Cases that don't respond to these common treatments are often referred to as intractable or treatment-resistant. This is unfortunately a rather high fraction of the total. Nobody knows the exact percentage, but many psychiatrists estimate it to be at least 20-30 percent of patients diagnosed with depression.

For severe and intractable depression, desperation once again comes into play. There is even a treatment for depression called electroconvulsive therapy, or ECT for short, in which physicians induce a seizure. They do this by pasting large electrodes to the patient's head and turning up the current. It's controversial and ugly, but it often works—and while nobody understands exactly why it works, it seems to require a seizure. I'll have more to say about ECT in the next chapter.

In a small but growing number of cases, patients with treatment-resistant depression are allowing physicians to implant electrodes in their brains. So far there have been only a handful of trials, but if they're successful then there could be a boom. It's certainly attracted a lot of media attention these days; as of 2012 I've seen stories about DBS and depression in such established venues as *Scientific American*, CNN, *Psychology Today*, *Wall Street Journal*, CBS News, and many others. The reports vary—a mixture of hype and realism, of gushing optimism as well as the more cautious variety of optimism.

To employ DBS, physicians and patients must address an

important question first: Where in the brain do you stick the electrodes? Which structure should you stimulate?

With movement disorders, this decision was informed by the success of certain ablation surgeries as well as some knowledge of the brain circuitry that seemed to be most affected in the disease. The same was basically true of OCD.

Depression is different. Researchers studying the physiological manifestations of this disorder have tended to focus on neurotransmitter systems rather than specific parts of the brain. Depression is hypothesized to be a problem with messaging, and in particular, of the message carried by a specific neurotransmitter, or set of neurotransmitters. Early studies suggested an imbalance in the neurotransmitter dopamine, or a related neurotransmitter called norepinephrine—or both. Then Prozac came along, causing many researchers to revise their focus—Prozac appears to affect certain mechanisms of yet another neurotransmitter called serotonin. These days the theories have gotten more complicated, while continuing to explain little.

But if you want to look at the neuroanatomy of depression, a good place to start is with the nuclei containing neurons that use these transmitters. A subset of scientists have been doing this for some time, exploring animal models of depression as well as human patients.[13] A common technique to study human brain function is to image the brain. Common techniques include fMRI and PET. The use of imaging allows researchers to compare the brain activity of clinically diagnosed depression patients with that of people who are free of the disorder. Areas of abnormally high or low activity are suspect.

Still, with all that research, the physiology of depression remains a muddle. Lots of regions may be involved, as are lots of neurotransmitters, and even strange processes such as neurogenesis—the recently discovered ability of the human brain to generate new neurons. The safest bet is that depression is a mixture of things, probably a whole class of diseases that are lumped into the same category because we don't understand the brain well enough to distinguish one disease from another.[14]

Similarity of symptoms doesn't necessarily imply similarity of causes.

The bottom line: anyone who wants to use DBS for depression has to make some educated guesses about where to put the electrodes. You can also look back at the work of Heath and Delgado, but again, that's not too popular these days. Many people who tout DBS probably *cringe* when they hear those names.

As of 2012, there have basically been three educated guesses: ventral capsule/ventral striatum, nucleus accumbens, and area 25.

The vc/vs we've met before—it is a target of OCD as well. And that has a lot to do with why it has become a target for depression. In addition to the sophisticated science and research techniques mentioned above, physicians have used other clues to select stimulation sites. A big clue came when some OCD patients reported improvements in mood when their ventral cap-sule/ventral striatum was stimulated. So it was a no-brainer, so to speak, to test this area for a mood disorder such as depression.

Donald Malone and his colleagues at the Cleveland Clinic along with physicians at Brown Medical School and Massa-chusetts General Hospital have conducted a trial for vc/vs stimulation in depression. In 2009 they reported the results in 15 patients, and Malone summarized the data and added the results for two additional patients in a 2010 paper.[15] The average age was 46 and patients were currently taking antidepressants, and most of them had received ECT, all to little avail. After bilateral implantation of the electrodes, the high-frequency stimulation began. Out of the 17 patients, 9 of them (53 percent) had a favorable response after 3 months, as assessed by a rating scale frequently used by psychiatrists. Of those 9 patients, 6 improved so much that their depression was considered to be in remission. These percentages fluctuated during the course of the study, but at last follow-up, which occurred at different times for the patients (average was a little over 3 years after implantation), 12 of 17 (71 percent) were improved and 6 of those were in remission. These results are quite good. Remember, these were treatment-*resistant* patients.

Another target, the nucleus accumbens, belongs to the ventral striatum, so it's extremely close to the vc/vs site. The nucleus accumbens is sometimes referred to as a "pleasure center" a la James Olds because it is known to be involved in reinforcement and reward. In human imaging studies, the nucleus accumbens has been shown to be active while people are experiencing some forms of pleasure. It seems like common sense to stimulate a "pleasure center" to treat depression, although common-sense solutions don't always work in brain science. Recall that re-searchers don't even know what high-frequency stimulation does to a nucleus, and some of the effects appear to be inhibitory. And the stimulation studies of Heath cast some doubt on the whole interpretation of pleasurable stimulation—see chapters 5 and 6.

But trials conducted in Bonn and Cologne in Germany by Thomas Schlaepfer, Bettina Bewernick, and their colleagues have demonstrated that the efficacy of nucleus accumbens stimulation for treatment of depression is similar to vc/vs. In one study they found half of their 10 patients showed at least a 50 percent reduction in a certain scale of depression after 12 months.[16] In a four-year follow-up of a group of 11 patients the researchers reported that most of the improvement had been sustained.[17]

The third target I'll mention is area 25, so-called because it is number 25 of the 52 areas of the cerebral cortex identified by the German neurologist Korbinian Brodmann in the early 20th century.[18] Area 25 is located in the frontal region on the medial side, meaning it is facing the midline. (You can't see it in a traditional view of the lateral side of a hemisphere; you have to cut the brain in half and peek at the inner side—medial, that is—of a hemisphere.) Another name for this area is subcallosal cingulate gyrus.[19]

Area 25, along with the rest of the cingulate cortex, is involved in processing emotional information, and human imaging studies have found that area 25 is abnormally active in patients with depression. Therefore, if the parameters of high-frequency stimulation are chosen to create as much inhibition as possible—a temporary "lesion" effect, similar to the stimulation in PD and OCD—then DBS can be expected to help.

Several groups have tested this hypothesis, and in the process have helped a number of people with treatment-resistant depression. One group at the University Health Network in Toronto, Canada—Andres Lozano, Sidney Kennedy, and their colleagues—have reported success similar to or slightly better than DBS in other regions. A team at Emory University, led by Helen Mayberg, collaborated with the Toronto team on early studies, and then performed one of their own which included bipolar disorder—patients who experience periods of mania or hyperactivity in addition to periods of depression. The Emory researchers reported an excellent remission rate of 58 percent at the two-year point of stimulation.[20]

As with stimulating the motor cortex for pain relief, categorizing area 25 stimulation as "deep" brain stimulation seems a misnomer. It is, however, deep in the sense that it requires penetration of the cranium, and for area 25, you have to go down quite a distance because this part of the cortex is tucked away in the medial part of the brain. Another reason people refer to it as DBS is that it uses the same technology and procedures. And if you're particularly interested in promoting area 25 stimulation, it doesn't hurt if you can tie it in with a type of treatment that's had some success and has been around for a while.

St. Jude Medical, which is trying to muscle in on Medtronic's business, believes in area 25 stimulation so much that they are conducting a large study called BROADEN, which stands for Brodmann Area 25 Deep Brain Neuromodulation (they've even trademarked this acronym). Success in this and similar trials will be crucial in obtaining FDA approval for any device St. Jude Medical wants to market. The FDA can be hard to convince but the data from early trials look good, and a humanitarian exemption, while limiting the number of patients—and potential sales—is a likely goal.

* * *

Epilepsy is a disorder you might think would be amenable to DBS treatment. If you can switch off or inhibit a specific region of the brain with high-frequency stimulation, it would seem a simple matter to use DBS to put the brakes on the runaway excitation that occurs during seizures. Fight fire with fire, like firefighters who burn a ring around a wild fire, which uses up all the fuel so that the blaze can't pass beyond it.

But it's not that easy when dealing with the brain. It never is.

While the seizures of some patients seem to begin in a localized region—which Penfield and other surgeons took advantage of when they identified the region and removed it—it's difficult to find the right spot, and besides, the seizures may not be stoppable with stimulation. Widespread tissue damage may have occurred because of previous seizures, and stimulation from a few electrodes, even if they're in the middle of the bad area, may not help a lot. You'll also need to stimulate all the time, which may lead to adverse side effects, or you'll need the technology to recognize the initial stage of a seizure so that the device can apply electrical current at the right time—and that's a tough task.

The seizures of other patients appear to begin in vast regions of the brain at the same time. How are you going to stop that kind of widespread activity?

One possible way to put a damper on seizures is to provide some measure of general inhibition. That's what antiepileptic medications do, and there are a lot of them on the market these days. But for the 25 percent or so of patients who fail to respond to drug therapy, neurologists have to think about other options. If you're going to stop epilepsy with stimulation, you'd like to find a site or target that has a lot of connections and therefore potentially has an influence over a large portion of the brain. One such target is the vagus nerve.

The vagus nerve is a cranial nerve—the tenth one, in the standard numbering system. Named after a Latin word meaning wandering, the vagus nerve is thick and long, snaking through the chest and abdomen. Since it carries so much information and makes a lot of connections in the brain, researchers wondered if stimulating it could have widespread effects. This proved to be true, and in 1997 the FDA approved a vagus nerve stimulator, manufactured by Cyberonics, as a treatment for certain types of epilepsy. The stimulator is generally implanted in the chest, as with DBS, but no craniotomy is required because the electrodes are placed in the neck and wrap around the vagus nerve. Cyberonics likes to call it a "brain pacemaker." The stimulator turns itself on and off, generating pulses that last an adjustable amount of time. Vagus nerve stimulation generally doesn't eliminate seizures, but many patients experience a significant reduction in frequency and severity of their seizures.

As you've probably learned to expect by now, neuroscientists do not know why this works. If you ask, the explanations usually involve vague notions that the stimulation may activate certain biochemical pathways or may act to reset abnormal rhythms.

Because of the vagus nerve's widespread connections, researchers have hypothesized that stimulation could potentially treat a number of other disorders. In 2005 the FDA approved the Cyberonics device for use in cases of treatment-resistant depression.

But it isn't DBS. Researchers interested in deep brain stimulation have been looking for a nucleus with a lot of connections throughout the brain, similar to the vagus nerve. When you're searching for something with a lot of connections, a good place to start is anything that acts as a relay or switching station, sort of like an old telephone switchboard. There's nothing exactly like that in the brain, but a certain collection of nuclei that we've encountered many times in this book has a lot of relaying/processing functions—the thalamus, which routes a ton of information back and forth between the cerebral cortex and other neural structures. DBS researchers have focused on a particular nucleus called the anterior nucleus of the thalamus.

Trials with a Medtronic device have found that stimulation of the anterior nucleus of the thalamus reduces seizure activity. The results have convinced European authorities to approve the device, and in 2010 an FDA advisory panel also voted to recommend approval—but it was a lukewarm recommendation, which means the data from the trials didn't overwhelmingly impress the panel. Advisory panels consist of experts invited to express their opinions, but the FDA isn't obliged to follow their recommendations. As of July 2012, the FDA has yet to give their approval, and may demand another trial if they're not satisfied with the data thus far.

Other prospective applications of DBS are in extremely early stages of investigation. Thanks to the incomplete knowledge of DBS mechanisms, there's a long list of possibilities. Which, as I mentioned earlier, can be a good thing. Or a bad thing, which is sometimes the result of having too much of a good thing.

Researchers are considering using DBS to treat Tourette's syndrome—a disorder causing tics and other involuntary movements, occasionally accompanied with salty language—as well as Alzheimer's disease, addictive behaviors, various forms of brain damage, and behavioral disorders. All of these are worthy of discussion but I want to dwell on two of the more interesting ones.

One is coma. Severe brain damage can produce a range of states of altered consciousness, from basically unconscious—deep coma—to a vegetative state (with irregular periods of eye movements) and then what is called a minimally conscious state in which the patient occasionally shows signs of awareness but is clearly not fully functional. In 2007 Nicholas Schiff at Weill Cornell Medical College and his colleagues stimulated the central thalamus of a patient in a minimally conscious state. The result was a somewhat miraculous recovery. The patient was a 38-year-old man who had been assaulted five years earlier and had suffered a severe brain injury. After six months of DBS, some of the man's behavioral and cognitive skills gradually returned. He could feed himself and managed to talk.[21]

It was definitely a feel-good story, and received lavish attention from the media. "Reawakening the Dormant Mind," mused *Discover*.

"Reawakening" a "dormant mind" with brain stimulation is

actually an old idea, dating back to the go-go era of the 1960s. It's yet another one of those appealing notions that sound so logical. Let's suppose you have a patient whose brain has shut down because of trauma or some other reason. There's little or no brain activity registering on the EEG, for example. How do you treat him? Well, why don't you juice up his brain with electricity and see if you can kick-start it into gear?

The central thalamus is a good choice to stimulate. It seems to form part of the gateway to consciousness, receiving inputs from brainstem arousal centers—the regions of the brain that wake you up and create an alert state—and other parts of the brain involved in regulating the level of arousal. Human imaging studies have shown that the central thalamus is active during tasks requiring attention and concentration. The central thalamus will be a likely target for future DBS trials on minimally conscious patients.

But I wouldn't bet on many more miracles. Perhaps the biggest problem is that severe brain injuries can vary widely in specific sites and causes. This means that not all of them are going to respond in the same way, just like other complicated diseases or disabilities such as depression that probably have a number of different etiologies even though they are all classified as the same disorder. And DBS is unlikely to work for patients who are deeply comatose.

In most hospitals the typical treatment for patients in a coma is simply to provide essential nutrients and life-support and then sit back and hope for spontaneous recovery. Every once in a while, patients recover on their own. Sadly, many don't. DBS offers an active approach rather than a passive, hope-for-the-best attitude, and in this respect it would seem superior. But I have my doubts it will work very often. In many cases it could end up being a waste of time and money, and a real disappointment to family members who may have entertained high expectations. The best that can be hoped for in my opinion is that further research identifies a subset of patients who have a better than average chance of responding favorably to DBS.

Another DBS application about which I have an even stronger

reservation is the treatment of obesity. In 2006 Lozano and his colleagues at Toronto stimulated the hypothalamus in an obese man, but the patient didn't lose much weight over the long term (the stimulation did seem to dredge up old memories, however, which made physicians who treat Alzheimer patients sit up and say, "Hmmm, I wonder..."). Another small study with a few patients conducted by Donald Whiting and his colleagues was more successful. One patient at the West Virginia University Hospital reported less hunger soon after stimulation of the hypothalamus, and the story got national attention when ABC News included it in a special on brain surgery for weight loss, which aired in 2009.

A government web site reported in 2012 that Ali Rezai, now at Ohio State University, will soon begin a trial titled "Deep Brain Stimulation for the Treatment of Obesity."[22] But the details are sketchy and I couldn't find anyone at Ohio State who wanted to talk to me about it.

It's certainly an interesting extension of DBS. Surgeons and psychiatrists have been tacitly following in the footsteps of Heath and Delgado for decades. Now they're going back even further in history and channeling Walter Hess (see chapter 5).

Some critics have labeled the use of DBS to treat obesity as "cosmetic neurology." I'm not sure the term is appropriate because these patients have serious problems. And animal studies going back to Hess's day have shown that food intake can be influenced by hypothalamic stimulation. A study published in 2012 tested DBS of the ventromedial hypothalamus in female Göttingen minipigs, whose diet, physiology, and brain structures resemble primates. The frequency of the stimulation was 50 hertz, which is relatively low for DBS. They found that stimulated animals experienced much less weight gain even when overeating the same amount as control animals that didn't receive stimulation.[23] The researchers suggest that DBS invoked an elevation in metabolism.

I suppose many people who deride these efforts may be worried that DBS treatments might in the future include healthy people looking to lose that stubborn five pounds on the hips or thighs. I'll have more to say in a later chapter about the if-this-goes-on possibilities.

What bothers me the most about using DBS to treat problems such as obesity is the lack of any obvious brain pathology or abnormality in these patients. There is definitely a problem—overeating, or overnutrition, as some clinicians like to put it—but this is behavioral, not necessarily physiological. Is every behavioral problem a brain disorder?

Answering yes to this question seems to be the trend in neurology and psychiatry these days, or at least it seems that way to me. For example, for quite some time I've heard psychiatrists repeat their mantra concerning addiction—it's a disease, they say. Abusing drugs is not a choice, according to these psychiatrists, but rather a disease that requires treatment.

I'm not so sure. I'll certainly acknowledge that the continual abuse of certain drugs *leads* to a brain state characterized by imbalances in neurotransmitters and other problems that looks very much like a disease—it's called withdrawal. Withdrawal cravings occur because the brain has adjusted so much to the constant presence of the drug that a sudden disappearance of it disrupts the new equilibrium. The brain's physiology has adapted to an environment that includes the drug and demands or craves it to maintain the status quo.

But what started the addict on the road to this dire situation in the first place? Were those poor decisions a result of a brain disease? Or were they just bad choices made by someone who should have known better?

Some researchers encourage the hope that we will soon be able discern brain disease and health with the aid of tools such as imaging the brain. But we're still having trouble solving these problems; we can make pretty pseudo-color images of brain activity but we can't decipher what we see because nobody knows how the brain actually works. What does an increase in the activity of this or that nucleus mean? It could mean lots of different things, depending on the state of the rest of the brain, the context of the activity, and so forth. We just don't know.

Until we do, researchers are mostly just guessing. It's true that in a few cases you can actually come to a firm conclusion rather than just make an educated guess; for example, you can look at an image of a brain undergoing a seizure and tell that there's definitely a physiological problem—part or all of the brain is hyperactive. But other "abnormalities" in imaging may simply represent variability.

For behavioral problems such as eating "disorders," who's to say which people have brain disorders and which people just lack will power? Or is there any difference between the two cases? If you tend to see every problem as a disease, whether it's addiction or overeating or picking your nose in public (which can cause acute embarrassment leading to anxiety and fear of public places and reclusion and a dramatically lower quality of life), then you never think of human spirit or will power. Instead you think of brains—and consequently people—as machines. To fix any problem all you have to do is follow certain instructions: FOR OVEREATING ADJUST THIS KNOB or STIMULATE THIS BRAIN REGION.

\* \* \*

How far can DBS go? How far *should* it go? In addition to the admittedly emotional argument I described above, there are several factors that ought to be considered.

An increasingly prevalent belief among researchers is that many brain disorders are caused by disrupted rhythms. This belief will likely foster the use of a tool such as DBS that may be able to return the activity to its normal state. I expect in the coming years that DBS will expand the number of electrodes it uses to more than just one or two. Two electrodes are needed for bilateral stimulation—remember, there are two of everything in the brain because it has two hemispheres—but if you want to stimulate other structures at the same time, you need more electrodes. It's almost inevitable that surgeons will start doing this (again, following Heath and Delgado).

Another factor is the ever-present placebo effect. Recall the improvement in patients receiving retinal implants sometimes extended beyond the area of implantation. Merely undergoing an operation seems to have a physiological or psychological effect.

How do you tell if the improvement seen in DBS is real or a placebo? Clearly the FDA thinks it's real, at least for the stimulators that they've approved. But how can they be sure? To test the placebo effect of drugs you can give a group of patients a "sugar pill" instead of the real medication and observe whether these control patients do just as well as the ones who got the drug. But for DBS you can hardly drill a hole in someone's head and insert electrodes and subject him to all of the serious risks associated with the operation and then *not do anything to treat him.* I realize that medical ethics is a big subject full of opposing viewpoints on what is and is not acceptable in the course of finding out which treatments are actually effective. As for me, I wouldn't want to tell a DBS patient something like this: "We didn't actually do anything except drill a hole in your head and accidentally give you an infection. You were a control."

To get around this issue, researchers have conducted trials in which they delayed stimulation in some of the patients. These are temporary controls in which the patients don't realize they're not yet receiving treatment, so you can compare their responses to those who receive stimulation sooner. But everybody eventually gets treatment. Researchers and regulators are satisfied that the responses to stimulation are higher than the controls, so DBS is doing something beyond giving patients a psychological lift. But these clever tricks don't preclude the possibility that some of the longer term effects in these patients are unrelated to the actual stimulation.

Another concern is money. Specifically, the making of it, and the conflicts of interest that sometimes occur while doing so. Companies such as Medtronic have a lot riding on clinical studies. It's an expensive investment, and you only get a return if the device is approved and you can sell it to a lot of patients.

I won't criticize these businesses for promoting their products and using terms such as "brain pacemaker" which make DBS seem less threatening. If the FDA does its job, nobody will be able to foist a useless device on the American public—and in general I think the FDA does its job, or at least takes it seriously.

But a potential problem arises when companies such as Medtronic and St. Jude Medical get too cozy with physicians and surgeons who are performing these clinical trials. The FDA may do a good job of analyzing data but their decisions are only as sound as the data. I'm not accusing anybody of anything, but when I hear that a particular business or corporation is sponsoring this or that clinical trial, which often happens these days, I always wonder just what such sponsorship entails. Even dedicated professionals are subject to a particular lapse of judgment called confirmation bias, in which everything is seen as confirming one's expectations—and if something doesn't, it's dismissed as aberrant, irrelevant, or unreliable, without probing too deeply into the reasons why (except that it failed to show what you wanted it to). One's bias has a way of being confirmed quite often when there's money riding on the outcome.

Assuming all the data and trials and analysis have been and will continue to be above reproach—and I've got no evidence to the contrary—I still wonder if DBS might have a limited lifecycle, at least in its current guise. Scientific research often advances unpredictably. Sometimes a researcher looking at the effects of one procedure finds something else that might do the same job just as well. And in addition to these serendipitous discoveries, there are plenty of medical and healthcare companies that make a living doing something other than DBS. Pharmaceutical com-panies, for example. These companies have an insatiable need for new drugs, and they hire armies of researchers and technicians who are continually sniffing around for new sources of revenue to replace the drugs for which the patent has expired. If DBS becomes popular enough to generate a lot of sales, drug makers might take notice and say, "Hmmm, I wonder if there's a pill that can do something like that."

And one day someone might find one.

# 9
# Field Effects: Stimulating the Brain without Opening the Skull

The previous chapters of this book described the development of methods that use electrodes to stimulate the nervous system. After an unfortunate beginning in the first human case—Roberts Bartholow's experiment on Mary Rafferty went poorly—researchers became more careful. They slowly acquired a greater knowledge of the brain and they designed sophisticated stimulators. Today physicians routinely implant electrodes and stimulate various parts of the human brain to treat an increasing number of disorders.

But still, procedures such as deep brain stimulation require a craniotomy, and this presents a problem. Not only are there risks of infection and brain damage, as described earlier, but there's also a psychological factor. Even though most of us these days are accustomed to amazing medical technologies and we (more or less) confidently submit to the recommendations of our healthcare providers, a craniotomy will make a lot of people hesitate. *They're going to drill a hole in my head.* It seems an egregious form of trespassing, a violation of something God or nature intended to remain untouched.

Not surprisingly, there have been plenty of efforts to develop techniques of stimulating the brain without having to remove parts of the skull beforehand. I've mostly avoided these transcranial methods in favor of describing the invasive craniotomy-and-electrode methods because in my view the invasive work has thus far produced results that are more useful, more controversial, and frankly more interesting. But this could soon change. The wave of the future might start leaning toward a subtle across-the-skull approach rather than a brute force punch-right-through-it. Hence, this chapter. And because of its great potential, the transcranial approach will also be featured in the book's final two chapters, in which I have the chutzpah to envision future pos-sibilities and to predict where brain stimulation may be going.

First, some historical perspective. Ever since the time of Galvani and Volta, in the late 18th and early 19th centuries, electricity has been applied to the body in the hopes that some-thing medically useful might happen. These early pursuits generally go by the name of electrotherapy.

Although medically suspect, the physics of it is reasonably sound. The brain and the rest of the body's tissues are subject to the same laws of physics as every other material. An electric field is a physicist's name for a region in which an electric charge will experience a force. In simple terms, electric fields are created by separation of charges. For instance, a capacitor consists of two metal plates separated by a small distance; when charged— connected to a battery, let's say—positive charges accumulate on one plate and negative on the others, hungrily wanting to embrace but unable to cross the gap. Instead they set up a field. Any charge placed in the field will rapidly skitter to one or the other plate if the charge can move—positive charges to the negatively charged plate and negative charges to the other plate, because in electricity opposites attract.

Put someone's head in an electric field and any mobile charge in it will move. The brain contains a lot of mobile charges, called ions, which are floating in solution. Applying an electric field to the head will therefore produce a current and stimulate the brain. But with a field you're stimulating a broad region of the brain rather than applying current locally through small electrodes.

Another law of physics is that currents flowing through a conductor create magnetic fields. A magnetic field is a little bit complicated, defined in physics by its action on electric charges, and it is created by electric currents (even an iron magnet—its magnetism is believed to be caused by the motion of charges in atoms). The tiny currents that flow through ion channels as the brain goes about its normal business will therefore create magnetic fields. These magnetic fields are minuscule, but sensitive instruments can pick them up—the process of recording them is known as magnetoencephalography (MEG).

A changing magnetic field generates an electric field.[1] This means that if you apply a changing magnetic field to a material in which charges can move (a conductor, in other words), you will produce a flow of charges—a current. This process is called induction. It works in the brain as it does anywhere else. You can therefore use induction to generate electric fields in the brain.

If you want to be less fancy, you can run electricity through the brain by constructing a simple electrical circuit that includes the head. As the current flows through the circuit it will naturally flow through the brain as well. To do this you have to put the head in electrical contact with the other components of the circuit, which means attaching electrodes to the skin. But this isn't very effective because dry skin isn't a good conductor of electricity. Moisten the skin, however, and it conducts pretty well. But keeping it moist is a problem, so in most cases the electrodes are pasted to the skin with some kind of conducting gel. The rest of the body's tissues are fair conductors except for bone, but electric current will cross bone too, though with a lot of resistive loss.

During the golden age of electrotherapy virtually every malady seemed like a good candidate for electric treatment, and every part of the body—and I mean every nook and cranny—made an inviting target. Electricity in general was in vogue, and in large cities you could find machines in drug stores or other places with two metallic handles and a coin-activated current. You put a coin in the slot, grabbed the handles—one hand on each—and when you spun one of the handles the circuit was completed, and you received a jolt of refreshing electricity. At the very least it would wake you up—if it didn't stop your heart.

Physicians and researchers back then had only a hazy idea of what electricity did to the body. An article published in 1873 in *The Popular Science Monthly* described a number of effects, including some on the circulatory system: "Experiment proves that, under certain conditions, the electric current contracts the vessels, and thus checks the flow of blood into the organs. Now, a great number of disorders are marked by too rapid a flow of blood, by what are known as congestions. Some forms of delirium and brain-excitement, as also many hallucinations of the different senses, are thus marked, and these are entirely cured by the application of the electric current to the head."[2] The author of this article also states that the brain is especially susceptible to such interventions: "No organ possesses a vascular system so delicate and complex as the brain's, nor is there any so sensitive to the action of causes that modify the circulation. For this reason, disorders seated in the brain are peculiarly amenable to electric treatment, and when carefully applied, it is remedial in brain-fevers, mental delirium, headaches, and sleeplessness."

By the late 19th century electrotherapy had reached its zenith. In 1901, physician and electrotherapy advocate William J. Herdman wrote that "about ten thousand physicians within the borders of the United States make use of electricity as a thera-peutical agent daily."[3] I have no idea how Herdman arrived at that number, nor do I know how many doctors were practicing at the time, but in 1906 the American Medical Association (AMA) published its first medical directory of licensed physicians in the U.S. and Canada, and the list contained more than 128,000 names.[4] Evidently only a fraction of doctors were using electrotherapy on a daily basis if Herdman's estimate is correct. But Herdman went on to note that many other physicians made use of electricity on occasion. Most of the medical professionals working at well-equipped hospitals would have certainly had access to the equip-ment, and progressive-minded physicians such as Roberts Bartholow would not have been shy about employing it. Indeed, doctors would have been deemed out-of-date if they didn't.

A common form of general-purpose electrotherapy was the electric bath.[5] The patient sat in a tub made of wood or ceramic filled with water. A large electrode was placed in the water at each end of the tub and the current was switched on. The bath solved the problem of skin conductivity, and at least some of the current flowing through the water entered the patient.

For more localized effects, the bathtub was ditched and doctors instead attached electrodes directly to the affected site. In the treatment known as galvanism—which used direct current, as from a battery—the brain was a common target. Electrodes were placed in various geometries, such as at the front and the back of the head, or ear-to-ear, or perhaps forehead-to-jaw. All sorts of ailments were treated. After the current was switched on the patients would feel their skin get warm, then they might start getting sleepy. If the physician increased the current, the result might be dizziness, the perception of light flashes, a metallic taste in the mouth, and eye movements. Still greater current could result in a seizure.

Application of alternating current, then known as faradic current, could also be applied. Guillaume Duchenne, a 19th century French neurologist, became well known for such "faradism."

Even at its height electrotherapy had its critics, and it faded shortly after the beginning of the 20th century. Some people believed that most of the positive effects of electrotherapy were all in the mind—a placebo effect—and others scoffed at electrotherapy because its advocates clearly didn't really know how electricity affected the body and brain. When Horsley, Cushing, and others began applying electricity to the exposed brain during surgery, as described in chapter 3, the attention of the medical community turned to invasive procedures.

Still, some electrotherapies have had staying power. X-rays, discovered in 1895, were originally classified as electrotherapy (and occasionally criticized and scorned as such). And, strange as it sounds, a dangerous side-effect of electrotherapy later began to find use in places such as hospitals for the mentally ill. It turns out that sometimes eliciting a seizure with electrical stimulation has unexpected benefits.

* * *

In the 1930s and 1940s, physicians and patients who were desperate to treat severe psychiatric disorders hit upon the technique of inducing a seizure by stimulating the brain with too much electricity. They pasted large electrodes to the head and turned up the current until the patient began convulsing, during which the patient writhed because the brain was firing wildly and sending impulses to the motor system, causing muscular spasms. Afterward, many patients appeared calmer. As a treatment it sounds bizarre but it's certainly not the only weird thing psychiatrists have tried that seems to work. Around the same time period physicians began treating severe mental illness with insulin comas. An insulin coma is a state of stupor or loss of consciousness induced by a large dose of insulin. The insulin rapidly lowers blood sugar (glucose), which is what the brain normally uses for energy, so the brain begins to shut down.[6]

Both of these treatments were originally used on patients with schizophrenia. Seizures and insulin comas tended to sedate the patients for a period of time after the treatment, making them easier to talk to and treat in other ways. Insulin coma has largely disappeared from the world's therapeutic repertoire, but seizure-inducing treatment remains, though it is used today mostly for the treatment of depression.

How does inducing a seizure alleviate the symptoms of depression? Perhaps it works by resetting the brain. Imagine a computer in which all the hardware switches and software settings have been scrambled because various users have fiddled with them over the years. Maybe things are so bad that the computer no longer functions. If you had a reset button you could restore the manufacturer's defaults—the configuration the computer had when you pulled it out of the box—and get the computer to work again.

Maybe this analogy has some validity, but I doubt it. The brain isn't a computer, and a seizure is more than just a reset button. Something much more complicated is at play here.

But empiricism rules in psychiatry. If it works, or seems to work, do it again, even if you haven't got a clue *how* it works. And this is what happened. More and more researchers adopted this technique, which became known as shock treatment or electroshock therapy. Many patients receiving the treatment reported feeling better, less depressed. Treatments have continued to the present day, even though the first-line of defense against depression these days generally involves drugs such as Prozac or Zoloft. But some patients fail to respond to drugs, and if their symptoms are severe enough, they're ready to try anything. Of course in today's era of euphemisms the term "shock treatment" just won't do. The modern term is electroconvulsive therapy, or ECT for short.

It's not only euphemistic it's downright inaccurate, since patients undergoing ECT these days are given muscle relaxants and anesthesia, so there is little convulsing going on. Anesthesia generally makes it more difficult to induce a seizure but a little additional voltage will take care of the problem. Physicians monitor the patient's brain activity with electrical recordings such as EEGs to determine when a seizure occurs.

The procedure has always been controversial. Some people find it barbaric, others call it lifesaving. Among the proponents are Kitty Dukakis, wife of 1988 Democratic presidential candidate Michael Dukakis. In her 2006 book, *Shock: The Healing Power of Electroconvulsive Therapy*, co-authored with Larry Tye, she describes her ordeal with depression and how ECT helped her to recover. Max Fink is another name I've long associated with advocating ECT—he's a psychiatrist who's been employing the technique and writing about it for decades. On the other hand, critics such as Peter Breggin—the long-time gadfly of psychiatry— and Linda Andre, a writer and activist, beg to differ on the general merits of this form of treatment.

One of the biggest problems with ECT is in public relations—it got a terrible reputation in the bad old days of psychiatry. Before the development of effective drugs, patients with severe psychiatric problems were housed in back wards of hospitals. Exasperated or frightened attendants may have commonly used treatments such as ECT as a form of punishment in an attempt to control violent or unruly patients. Without anesthesia and muscle relaxants, an electric shock strong enough to induce a seizure is an experience you'd probably want to avoid if you could help it. In addition to the seizure itself, excessive currents can burn the skin or cause intense pain by activating pain receptors in the face or scalp (recall there are no pain receptors in the brain, but there are pain receptors in the skin where the electrodes are attached). The 1975 movie *One Flew Over the Cuckoo's Nest,* which portrayed the abuse of these treatments by sadistic personnel in a psychiatric ward, is often cited as giving ECT a bad name.[7]

Countering that kind of negative publicity is difficult. Maybe the best indication that it's an effective treatment is that it still exists even after all that controversy. According to the estimates I've seen, about 80 percent or so of patients get at least some short-term relief from ECT. Patients usually receive treatments several times a week for 2-4 weeks, although in some cases the sessions will be stopped if the patients have a particularly rapid recovery. But over the longer-term—months and years after the sessions—the outcomes seem to be less impressive.

One of the most common side effects is memory loss, especially memories of things that happened a short time before treatment—an effect called retrograde amnesia. The seizures apparently interfere with the process of transferring memories from short-term storage to a more permanent form, a process known as consolidation. This means you tend to lose memories of recent events, which are held in a more fragile state than long-term memories.

The biggest unknown is how ECT works. About the only thing physicians know for certain about ECT is that it has to evoke an actual seizure for it to have any chance of working. A sub-threshold bit of electricity, even if substantial, doesn't do much.

Although the mechanism remains mysterious, many ECT practitioners and advocates seem confident, at least publicly, that the treatment doesn't involve brain damage. Even mention of the phrase is sometimes enough to rile a mild-mannered psychiatrist.

It wasn't always this way. I remember when I was a student back in the early 1990s listening to conversations among neurologists about where and how the damage associated with ECT occurs—not *if* there's damage, which was assumed, but its extent. These days it's more like: "Damage? Are you kidding? There's no damage!"

If you've read this far in the book, you're now aware that brain-damaging treatments for neurological and psychiatric disorders are fairly common. Recall that in addition to the controversial lobotomies, there are far less controversial treatments for epilepsy, Parkinson's disease, obsessive-compulsive disorder, and others involving ablation or destruction of certain parts of the brain. Brain damage in these treatments isn't just a side effect, it's actually part of the process. Although many patients these days have the option of deep brain stimulation, ablation treatments continue. And for good reason—they're often successful. It's obviously not something to be taken lightly, but if it's what you have to do, it seems better to lose a little brain tissue than deal with the symptoms of a debilitating disorder.

But brain damage seems to have been ruled out of the ECT equation, if you believe what many of its practitioners tell you. For example, Dukehealth.org, the web site of Duke University medical schools and hospitals, posted an interview concerning ECT with Dr. Sarah Hollingsworth Lisanby. When asked if ECT damages the brain, Dr. Lisanby replied, "No. Careful studies using sensitive brain imaging measures in people receiving ECT, and precise anatomical measurements in animal research studies, have repeatedly demonstrated that ECT does not damage the brain."

I admire physicians and healthcare institutions that try to educate their patients and cite evidence to back up their assertions. It's refreshing and commendable, and would make me feel much more comfortable as a patient under their care. But in this particular case, I have a deep suspicion that their assertion is incorrect.

Even the best imaging devices available today do not have the resolution to confirm a lack of damage. Damage need not consist of a massive loss of cells—it can be much more subtle than that, involving molecular or biochemical systems. Such damage isn't necessarily going to show up in "anatomical measurements" on animals either.

In my view it's highly unlikely you're going to induce

something as drastic as a seizure three times a week for nearly a month and not produce *some* damage. The brain is far too delicate for that. Don't tell me you're going to hit it repeatedly with the electrical equivalent of a sledgehammer and nothing is going to break—that's just too incredible for me to believe. Maybe you don't have to call it "damage," maybe you can call it permanent alteration, or seizure-induced morphological adaptation. But it's *something*. The existence of amnesiac side-effects clearly demonstrates you've created some kind of lasting change.

In 2012, Jennifer S. Perrin, Christian Schwarzbauer and their colleagues at the University of Aberdeen along with researchers at the University of Dundee got some attention for their paper on ECT published in the *Proceedings of the National Academy of Sciences*.[8] The researchers used functional MRI (fMRI) and a sophisticated statistical analysis to measure the functional connectivity of certain areas of the brain. Functional connectivity refers to correlations in the activity of brain regions by which researchers infer that these regions have an influence on each other—their functions are in some way related, and they may be communicating, for instance, through synapses. The researchers studied patients before and after treatment and found that functional connectivity decreased in some cortical regions after ECT. This fits with some previous studies that have suggested mood disorders may be associated with abnormally high connectivity.

I doubt it's the whole story, but it's an interesting development. It's important that such studies continue; if the mechanisms underlying ECT can be identified, maybe physicians will be able to implement the same effect without having to induce a seizure.

Or, in a slightly less ideal world, people will be more accepting of treatments that work but don't necessary make sense and can, in the wrong hands, be abusive. It's an uphill struggle. I'll grant the need for caution and openness, and I also appreciate the concern of ECT's many vocal critics. But it seems to me that some critics have reached a point where they care more about their agenda than the patients. If informed consent is given and the patients get the results they want—which in many cases they do—what's the problem? Maybe it's time for paternalistic critics to find something else to do besides trying to protect people from choices they willingly decide to make.

* * *

What happens when you paste electrodes to someone's head but don't crank up the current high enough to elicit a seizure? You won't generally get any relief from depression, but there are many other things low-level stimulation might achieve. And there's a broad range of current intensities you can safely try, from those that elicit a perceptible tingling to one so low it can't be perceived—a subliminal amount. What can these currents do?

Almost anything, it seems—if you believe certain advertisements or books and articles written by "experts." You can relax your mind, deepen your concentration, fall asleep or become alert, lose your craving for sweets or drugs or sex, overcome fears, relieve pain and tension, boost your memory, enhance your intelligence (or even better, your offspring's intelligence), you can even reach a transcendental plane of higher dimensions and be filled with the inner light of universal consciousness. Welcome to the new age of electrotherapeutics.

Actually, it never quite went away. But the 1960s brought it roaring back, along with the cultural revolution and what I've called the go-go era of science. It's hard to think of anything that someone somewhere hasn't claimed is possible when mild electric currents are applied to the head or other parts of the body.

The Soviets, it seems, were always fascinated with scientific research programs aimed at rather extravagant claims.[9] To cite an example, in 1966 *New Scientist* reported the opening of a clinic in London that provides "probably the most innocuous form of artificial sleep"—a treatment that is "widely practiced in Russia."[10] Electrodes attached to the forehead and a lower part of the skull deliver millisecond pulses of tiny currents. Developed in the Soviet Union, the machine puts you safely to sleep, although—and you might have guessed this—"the theoretical basis remains uncertain." The article also unsurprisingly claims it can work for a variety of problems: it helps in treating depression, anxiety, drug abuse, headache, schizophrenia, ulcers, high blood pressure, asthma, emphysema, morning sickness during pregnancy, as well as acting as an anesthetic.

Although all of this sounds too good to be true, electrical stimulation for the treatment of insomnia would seem to make sense. The brain rhythms that occur during the several stages of sleep are well known because researchers have recorded and studied them for a long time. So why can't we elicit or foster these sleep rhythms with electrical stimulation?

I suspect you know how I'm going to answer that question. The answer is, the brain isn't that simple. While replicating these rhythms would seem to be a great way of inducing sleep on command, you can't easily call them up. Though scientists have been making progress understanding sleep, it is still in many ways mysterious. It's a global state of the brain involving bio-chemistry and the interplay of several important regions of the brain, including the brainstem, thalamus, and cortex. Nobody is even certain why we sleep.[11]

Why does electrical stimulation seem to work? You probably know what my answer will be to this one as well—the placebo effect. But in some circumstances there might be more to it than that. Sleep-inducing electrical stimulation can actually work on some people if that's what it takes to relax them. Anything that relaxes you can get you to fall asleep. There's no miracle here.

Unlike invasive procedures involving craniotomies and neurosurgery and anesthesia, pasting electrodes on someone's head and turning on the stimulator is about as cheap and easy as scientific research can get. That's why research on transcranial stimulation has continued and will do so for a long time to come.

Gradually researchers began to test the therapeutic potential of this technique in earnest. In 1997, for example, researchers at the National Institute on Drug Abuse published a paper on the efficacy of transcranial stimulation—which they called cranial electrical stimulation (CES)—on tobacco withdrawal and smoking cessation.[12] The researchers placed electrodes on both sides of a subject's head (specifically, the mastoid process, located behind the ear). The stimulator delivered current pulses that were too small for the subjects to detect. This allowed the researchers to set up what is known as a double blind study. It involved two groups—the experimental group that received the treatment and a control group that didn't. Because the stimulation was imperceptible, the subjects didn't know which group they were in—they couldn't tell if they were actually getting the treatment or not, so they entered the study "blind." The selection of subjects into control and experimental groups was randomized and the researchers didn't know who was in what group, so they were "blind" too. Double-blind studies attempt to account for positive responses due to the placebo effect and also eliminate confirmation bias on the part of the researchers.

The study was a large one, with 51 smokers in the experimental group and 50 in the control group. For 5 consecutive days the experimental subjects received stimulation (and the control group wore the electrodes but the current wasn't switched on). The findings were disappointing: the researchers found no significant differences between the experimental and control groups on a variety of smoking measures, including number of cigarettes smoked and cravings. The experimental group did report less craving and anxiety on the first 2 days of the test, however.

Researchers at the University of Freiburg in Germany published in 1998 the results of a test of what they called "mind machines."[13] These devices include subliminal electrical stimulators as well as machines that provide low-intensity sensory stimuli such as flickering lights and slowly modulated sounds. The idea is to modify the user's brain rhythms. Manufacturers of these devices claim that in relaxation mode they can generate alpha waves; these are waves of about 8-12 hertz that are commonly recorded on EEGs from people who are in a relaxed state. Some proponents claim that altered or transcendental states of consciousness can also be invoked.

The study measured the responses of 30 volunteers to each of four conditions: resting, sham stimulation, transcranial stimulation, and sensory stimulation (optico-acoustic). Responses included questionnaires designed to reveal the subjects' psychological state and the measurement of skin resistance, a sign of anxiety. The researchers found no significant differences in the skin resistance measurements among the four conditions. Anxiety and tension, as measured by questionnaires, dropped by a large factor in all conditions—which meant the treatment worked, but it was the placebo effect, because doing nothing also worked just as well. However, the researchers found that electrical and sensory stimulation "produced significantly more visionary experiences and fear of ego dissolution than rest and placebo." Apparently the stimulation did something, as revealed in the responses to the psychological tests, perhaps enough to set off some subconscious processes.

Electrical stimulation seems to have fallen out of favor these days among the alternative therapy crowd. What I found recently when searching for "mind machines" on the Internet is a lot of sensory stimulators—light and sound—though there are still a few electrical stimulators on the market. None of these things are endorsed by medical authorities or government agencies.

Perhaps subliminal or ultralow-intensity transcranial stimulation could do more if you could find the right parameters, such as frequency, the shape of the pulses, position of the electrodes, and so on. But it seems unlikely to me. People who promote the use of these devices often present what seems to be a logically simple and straightforward argument: the brain "runs" on electricity, so if you provide a little extra current, maybe it will run better. But this argument is basically the same as saying that cars run on gasoline and therefore if you pour a little gasoline on the hood the car will run better. The brain didn't evolve to respond to low-level stimulation and it's not at all clear to me how such stimulation can exert any sort of benefit.

And yet, since we don't know very much about how brains work, I have to admit anything is possible. Interest has recently been renewed in the clinical application of low-level electrical stimulation called transcranial direct current stimulation, which I will describe in a later section in this chapter.

* * *

Current passing through electrodes pasted on the head can have a powerful effect if the current is strong, as in ECT, or little or no effect if the current is extremely weak. But what about the energy fields that continually surround us these days? Our bodies are constantly bombarded with electromagnetic energy coming from transmitters, power lines, and electronic gadgets of all kinds. We live in a digital world, full of electrical devices. Some people are concerned that this inadvertent stimulation may have a serious impact on our health and our minds.

Biologists and biomedical agencies often use the term "electromagnetic field" or EMF when discussing this type of stimulation. I described electric and magnetic fields earlier. Since electricity and magnetism are related, these phenomena are often collectively referred to as electromagnetism. Electrical circuits produce electric and magnetic fields, and these fields can extend for some distance from the circuit. They can also propagate—a changing magnetic field produces an electric field, which in turn generates a magnetic field, so these fields can go on generating one another and travel through space as well as through matter. These propagating fields are called electromagnetic radiation or waves.[14] Circuits specifically designed to emit a lot of electromagnetic radiation at a certain frequency will usually contain an antenna, though some emission generally occurs even without one. Researchers who study the biological effects of EMF may use this term to refer to electric and magnetic fields close to the device or circuit, but the term may also refer to the emitted radiation as well.

If you're close to an electrical circuit or device, you may be subjected to its electric and magnetic fields. A commonly cited example is a power line. Although in many cities the wires that carry electricity to homes are buried rather than strung overhead, wires carrying electricity for long distances along the power grid are often above the surface. These power lines have extremely high voltage, usually thousands of volts. (High voltage transmission is more efficient, resulting in less resistive loss. A low voltage would dissipate before it reached its destination.) Because of this high voltage, power lines have substantial electric and magnetic fields.

Even if you're far away from the circuit, you still might be affected by its energy if it's a powerful emitter of radiation. Electromagnetic radiation comes in a range of frequencies, from high-frequency X-rays down to relatively low-frequency radio waves and microwaves. Light is somewhere in the middle in terms of frequency. The higher the frequency, the greater the energy. The radio waves used to carry television and other broadcast transmissions have a frequency of around a few hundred thousand to a few hundred million hertz. Power lines also radiate, but their frequency is extremely low—60 hertz in the United States—and most people aren't too worried about these emissions.

The subject of EMF is contentious, but it's important to understand why. Nobody disputes the danger of strong electromagnetic fields. They can disrupt brain function, they can burn you, cause cancer, and so on. The controversy arises when you ask how much EMF is too much. Does weak EMF have deleterious effects? If so, how bad?

This book is about electrical stimulation of the brain, and since EMF has been alleged to affect the brain, I've included a discussion of it in this chapter. But you should be aware that this is only a small part of the EMF story. People have attributed all sorts of damage to EMF, and the concerns have been around for a while—as a kid in the 1960s I remember my parents strenuously warning me not to get too close to our new color television set. (I was appalled when I saw my cousin watching TV with his head practically resting against the tube.) Weak EMF has been alleged to interfere with our physiological adaptation to Earth's magnetic field, and it has been reported to spur abnormal growth, induce genetic mutations, and much more.

A researcher who figured prominently in chapter 6 of this book also got involved in EMF research later in his career. After he moved to Spain in the 1970s, José M.R. Delgado grew interested in EMF because it was a way of stimulating the brain without the need for craniotomies. In the course of his studies he made some disturbing discoveries. For instance, chick embryos exposed to certain weak fields became severely deformed.

The findings of Delgado and other researchers such as Robert O. Becker were disconcerting because according to the conventional viewpoint, the only danger from weak-to-moderate EMF is thermal—it heats biological tissues. You've made use of this effect often if you own a microwave oven. Microwaves cook food from the inside; at certain frequencies (many microwave ovens operate at 2.45 billion hertz) electromagnetic radiation sets water and certain organic molecules into vibration and generates heat. But Delgado, Becker, and others claimed weak EMF can do much more than this.

How close is too close? It depends on various factors such as frequency, but a good thing about radiation is that its strength rapidly diminishes over distance. The energy density of radiation falls by the square of the distance (this principle is known as the inverse square law); if you increase your distance from the source by a factor of 10, you're exposed to $10^2 = 100$ times less energy.[15] A little bit of distance puts you safely out of the danger zone of everything except extremely powerful emitters.

Of course if you put the emitter next to your ear you might be a little too close. This is why some people are concerned about cell phones, which emit radiation in the microwave range while in communication with cell phone towers. Some research has suggested a link between cell phone usage and brain tumors, while other studies have found nothing.

This is typical of this kind of research. Ever since the warnings of Delgado and Becker in the 1970s and 1980s, a lot of research has been done. Data from some studies support the dangers of weak EMF, while other researchers claim it's harmless. The standard response these days from mainstream scientists and regulatory agencies—the "experts"—is that the levels you experience on a daily basis is probably safe, though some people may add the tired old caveat "more research is necessary."

I'm always leery of any kind of research having to do with environmental issues. Many of the scientists who study environmental issues have a clear agenda. Some of them seem determined to downplay any environmental effects of technology or industry, and other researchers are just as determined to set off the alarm bells. The results of these studies invariably confirm what the researchers already believe, and what they believe depends on their political orientation and ties to, or prejudices against, corporations and businesses. Confirmation bias rages like wildfire in environmental research. It's extremely difficult to evaluate these findings. I'm not even going to try.

But what about the brain and EMF? Interesting effects have been reported, some of which I'm willing to accept or at least regard as likely, and some of which are absurd.

Some researchers have claimed to entrain brain rhythms with low levels of EMF. Researchers have also found that EMF can generate blips called evoked potentials in a person's EEG; evoked potentials indicate that a stimulus had an effect on the brain and was processed, though this does not necessarily mean the person was conscious of it. These and other effects are interesting but it's hard to know what to make of them. But that doesn't stop people from making *something* out of them. And a lot of the claims are outrageous. One of these silly claims has been especially persistent, and dear to the heart of conspiracy theorists: EMF is used in mind control.

As with most scientific manias, the Soviets were "leaders" in the field. During the cold war, fear of the Soviets was so rampant that even normally level-headed people started worrying about the nefarious intentions of such odd research. News broke in 1976 that the Soviets were bombarding the U.S. embassy in Moscow with microwave radiation, a discovery that the American government had made in the 1960s but kept quiet. And the hubbub began, with then Senator Bob Dole, among others, calling for hearings.

The intensity of the radiation was extremely low, on the order of a few microwatts (a thousandth of the power of an average light bulb).[16] Johns Hopkins University conducted a study of the health of American personnel who served at the embassy during these years and reported no adverse health effects, although this conclusion remains a matter of some debate.

To my knowledge, the purpose of this radiation has never been revealed, but the main theories are that the Soviets were trying to jam American intelligence-gathering devices, or they were communicating with their own devices inside the U.S. embassy, or they were trying to influence the health or behavior of American personnel who lived and worked there. (Influence their behavior to do what? Give out secrets? Sign treaties favorable to the Soviet Union? Be more receptive to communist propaganda?)

I'm not sure what anyone could hope to accomplish with such a weak stimulus, but mind control wouldn't have even been conceivably possible. Even Delgado dismissed that theory. In an interview published in a 1985 issue of *Omni*, he is quoted as saying he didn't believe it.[17] He went on to say that the strength of the signal drops by too much for it to have worked (essentially he's citing the inverse square law). In my view, the main reason it's not possible is that the brain isn't that simple. You can't aim an incredibly weak beam of radiation at the brain and expect to get a meaningful response. Information is processed in too many subtle and complicated ways for the brain to be easily influenced, much less controlled, by some random form of irradiation.

With so much silliness around it's hard to take a topic like this seriously, but I think it would be a mistake to dismiss it entirely. Even fields with strengths too low to electrically excite neurons could possibly have some biochemical effects by moving ions around or altering protein configurations. It's conceivable that this could have a slight impact over the long term, but I don't know what, if anything, it might do.

* * *

A technique that indisputably has an effect on the brain is the use of relatively strong magnetic fields. A time-varying magnetic field applied to the head passes easily through the cranium and induces electric fields and currents in the brain. This technique is known as transcranial magnetic stimulation (TMS).

Scientists and physicians in the late 19th and early 20th centuries played around with magnets and used them to stimulate the brain, but these researchers didn't do much except make themselves dizzy or elicit perceptions of spots of light—phosphenes, which in this case could be called magneto-phosphenes, since they were invoked with magnetism. In the go-go era of the 1960s, scientists began exploring the use of pulsed magnetic fields to stimulate nerves in animals—an updated Galvani experiment, making frog legs twitch with magnetism.

But you need a really strong magnet to do this. All of us are constantly exposed to at least one magnetic field—the one our planet generates—but Earth's magnetic field is weak. It can move compass needles but it can't get a human neuron to fire an action potential.[18] To stimulate a nervous system you need a magnetic strength of a few Tesla, which is about 40,000 times greater than Earth's magnetic field at the surface of the planet and several hundred times the strength of those magnets people stick to the door of their refrigerator. It's not easy to construct and use a magnet of this strength, and of course there are safety concerns about putting one near someone's body. Physicians became interested in developing a magnetic stimulator to test nerve function as part of the diagnosis of certain diseases, but it wasn't until the 1970s that Anthony Barker and his colleagues at the University of Sheffield managed to build a reliable machine to deliver high-strength magnetic pulses to nerves. In 1985 Barker unveiled a device to stimulate the brain—the first transcranial magnetic stimulator.

A single pulse of TMS creates a rapid rise and fall of the magnetic field, a change that induces electric fields in nearby conductors (such as brain tissue). This pulse lasts only an extremely short period of time, and thus the field induced in the brain goes away in a fraction of a second. A series or train of pulses, however, can affect the brain well after the stimulation stops; this isn't a physical effect of the fields, which are no longer present after the stimulation ends, but rather a physiological effect—the brain's activity has been changed, at least for a while. The amount of time depends on how long the stimulus was applied, with the effect generally lasting about twice as long as the stimulation.[19] Stimulating with a series of pulses is called repetitive TMS, or rTMS. Using a series of pulses instead of a single pulse is common, and so sometimes people drop the "r" and refer to the repetitive version as just TMS.

The TMS machine consists of a coil in the shape of a circle or figure eight and looks like a small paddle. The coil is placed on the head over the brain region to be stimulated. A stimulator or generator is attached to the coil. Electric currents flowing through the coil generate strong magnetic fields. I've never had my brain magnetically stimulated but in the early days of these devices I once let a colleague use one on my arm. It tingled.

The cranium does little to deflect or diminish the field. Inside the brain, the time-varying magnetic fields induce electric fields which drive ions, thereby producing current. The amount of current depends on the parameters of the stimulation, such as the rate of change of the magnetic field. This stimulation only penetrates about a centimeter (0.4 inches) or two in the brain and covers about the same width, for a total volume of about one or so cubic centimeters of brain tissue. So we're talking mostly cortex here, plus some of the fiber tracts—the white matter, or axons, underneath—plus maybe the upper portion of a few of the shallow subcortical structures if you place the coil in just the right spot. But cortex is generally always the target.

You might think the term "stimulation" in TMS would imply that the effects are excitatory, stimulating neurons into action. But recall DBS and its complicated effects, some of which are inhibitory and reduce or eliminate neural firing. It's the same case with TMS. The net effect depends in part on the frequency of the pulses.

In the 1990s, TMS became popular with researchers as a relatively affordable technique of studying the brain by stimulating it. (Imaging the brain with scanners such as fMRI is expensive because these machines can cost several million dollars, whereas TMS machines cost about $30,000.) No one knew exactly what the stimulation was doing, but in most cases it seemed to disrupt the activity and proper functioning of the stimulated cortical site. Which makes sense—as in other cases of brain stimulation, you're gumming up the works. Just as you can invoke confusion or arrest ongoing activity in a person by electrically stimulating the brain, TMS often temporarily disables a small region of cortex by suppressing the activity or by inserting a lot of signals that don't belong there. In other words, TMS may inhibit an area or introduce a lot of noise that drowns out the normal activity. Researchers were delighted to have a technique that allowed them to disable or effectively remove a specific cortical region for a short period of time and study the effects. It was called a virtual lesion—it mimicked a brain injury, though it was transient and harmless.[20]

Along with imaging studies, TMS has helped neuroscientists construct a functional map of the human cortex. Although the details are fuzzy, researchers now have a better idea which part of the cortex does what, at least for relatively simple functions. But there's still a lot we don't know, especially how different areas work together to perform complex operations and create mysterious processes such as consciousness.

But as time went on, scientists realized that the simple virtual lesion idea was, well, perhaps a little too simple. Some researchers have discovered an enhancement effect when certain cortical areas are "virtually lesioned." It's not all that hard to understand—consider the improvement in patient symptoms occasioned in some cases by ablation—but still, enhancement of a function doesn't quite fit with the description of the effect as an injury. Carlo Miniussi at the University of Brescia in Italy and his colleagues noted as much in 2010 when they wrote, "Interpretations of data can go beyond a simple relationship between an anatomical area and impairment of behaviour as suggested by the virtual lesion terminology."[21]

It bears repeating—I've repeated it often enough in this book, but it's a fine mantra, and perhaps the only absolute truth we've discovered in neuroscience—the brain is *never* that simple. I'll take up the issue of TMS-induced enhancement in the following chapter.

Since one of the motivations for the development of TMS was its use in a clinical setting, it's not surprising that researchers quickly began exploring the effects of magnetic stimulation on patients with disorders such as depression. Perhaps you could relieve some of the patient's symptoms if you could disable a malfunctioning region of cortex. Or perhaps you could activate it with excitatory stimulation if it is too quiet.

Researchers quickly discovered that in certain cases such as drug-resistant depression TMS is beneficial. The targeted area is a part of the frontal cortex on the left side that has a tendency to be relatively inactive in depressed patients, as indicated in imaging studies. Although the exact physiological actions of TMS are still uncertain, physicians adjusted the stimulus parameters in an effort to stimulate the activity of this region of the left frontal lobe. Clinical trials went well. In 2008 the FDA approved a TMS device made by Neuronetics, a company headquartered in Malvern, Pennsylvania. (As usual, other countries were well ahead of the United States. Canadian medical authorities approved TMS for depression in 2002.) The device is called NeuroStar. Patients sit in what looks like a dentist chair, with the coil on a boom that swivels into place. The dentist chair would be enough to make me feel uncomfortable, but the TMS procedure is painless. Treatments last about 40 minutes and are given every weekday for about a month or so.

Since the patients don't feel anything, it was easy for physicians who conducted the clinical trials to test for the placebo effect (patients weren't aware if the device was switched on or not). The treatment delivered better results than a placebo, so the FDA's decision to approve TMS was an easy one. About half of patients undergoing TMS experience a significant reduction in their symptoms. So it doesn't work for everybody, but these are some of the hardest cases. Another selling point for TMS is that it has few side effects. A small percentage—about five percent or so—of patients suffer from headaches or other pain and discontinue the treatment.

But anytime you inject or induce electric current in the brain, the possibility of a seizure exists. Seizures have occurred during repetitive TMS, though they are exceptionally rare. Unless, of course, you *want* to invoke a seizure, in which case you crank up the magnetic field or leave it on for a long time. Some progressive doctors have their eye on this very thing as an alternative to ECT. With TMS researchers hope to induce a seizure in much more limited way than ECT. Instead of a sledgehammer, you use a tool with a lighter touch. The hope is to achieve the same clinical benefit but minimize the side effects, such as memory loss, associated with ECT. Tests of this procedure, known as magnetic seizure therapy, are underway. Dr. Lisanby and her colleagues at Duke are among those studying the safety and efficacy of this approach.

Researchers are also looking into the use of TMS for a large number of other psychiatric and neurological disorders. Thus far, depression is the only success.

TMS is still a new method, and has a promising future. It has relieved the symptoms for many patients in which drug treatment fails to work or has unbearable side effects. But any time I see a new and somewhat mysterious treatment, I have to wonder about the long-term effects. It clearly has benefits for many patients, and in these people it must be doing *something* to the brain. Patients don't get better just because of the stimulation—the fields go away when the machine is turned off. So what's going on? The stimulation-induced electrical activity must indirectly have some kind of long-lasting effect. Perhaps it adjusts certain biochemical reactions. Or perhaps there is a change in gene expression, in which certain proteins are made from their genetic template. Do these effects attack the problem itself and perform at least a partial fix, or are these effects merely compensating for a problem that's still present? When we can answer such questions we'll make a lot more progress. And, like DBS, the possibility exists that we can dispense with TMS if someone can find a more efficient way of invoking these physiological mechanisms.

* * *

The development of TMS stimulated more things than just brains. The popularity of magnetic stimulation brought about a renewed interest in low-level electrical stimulation with electrodes. It's an idea that refuses to die. It has a fancy name now—transcranial direct current stimulation, or tDCS for short—but it's basically the same as what folks back in the old days called galvanic stimulation.

What advantages could it have over TMS? As I mentioned earlier, it's cheap—about the cheapest way you can stimulate the brain with electricity. It also may give you a little more flexibility in the way you configure the system.

These days the electrodes are often encased in sponges soaked in saline (salt water) to reduce the problem of dry-skin resistance. The current is generally one or two milliamps. Although almost any amount of electricity can be dangerous, a current of a few milliamps is pretty small. The stimulation can last a few minutes up to 30 minutes or sometimes even longer.

In studying how tDCS affects the brain, researchers today can employ far more sensitive instruments than in the past. They also have sophisticated computers and computer modeling software to help them understand how currents flow in the brain and how this affects neurons. Even though there is still uncertainty, scientists believe that weak DC does not cause neurons to fire action potentials. It does, however, influence the membrane potential in some way. Exactly how weak DC causes the membrane potential to change is the subject of some debate, but however it happens, it gives neurons a greater or lesser tendency to fire an action potential depending on which direction the membrane potential moves.

Weak DC's effect on membrane potential depends on several things, only some of which can be controlled. The orientation of the axons, dendrites, and cell bodies with respect to the currents and electric fields has a lot to do with whether they're going to be excited (depolarized) or not (hyperpolarized). There's not much you can do about how the cells and cell compartments are situated, though.

One thing you do have control over is the polarity of the electrode. A battery or any other source of direct-current electricity has two terminals—positive and negative—and the electrode that is hooked to the positive terminal is called the anode and the one hooked to the negative side is the cathode. Researchers have discovered that when you put the anode over certain cortical regions such as the motor cortex, this cortical region becomes more excitable. In other words, the cells become more liable to fire action potentials. The opposite is true when the cathode is placed over the motor cortex. Something similar happens when you put the cathode or anode over the visual cortex.[22]

So you have some options. In general, anodes excite and cathodes suppress. Jerome Brunelin at the University of Lyon in France and his colleagues published a paper in 2012 in which they made use of polarity in the placement of electrodes.[23] The researchers were interested in the abatement of auditory verbal hallucinations—hearing voices—in patients diagnosed with schizophrenia. In a substantial fraction of patients these hallucinations remain even after administering medication. Brunelin and his colleagues tested 30 of these patients with tDCS configured to excite a portion of the left frontal cortex called the dorsolateral prefrontal cortex[24] (DLPFC), and inhibit a region further back, around the border of the temporal and parietal cortices. In other words, they attached the anode over the left frontal cortex and the cathode over the left temporo-parietal cortex. The experimental group received 20 minutes of 2-milliamp stimulation two times a day for five consecutive days, while the control group got sham stimulation. The results were encouraging. Hallucinations were reduced in the experimental group compared to the control group, with an average reduction of 31 percent. The abatement lasted up to three months.

Researchers at the University of New South Wales in Australia and their colleagues conducted an even larger study on the merits of tDCS in treatment of drug-resistant depression.[25] As with TMS, the stimulation was configured to invigorate a certain region of the left prefrontal cortex (the anode was placed over this area). A total of 64 patients participated, and the experimental group received a daily 20-minute dose of tDCS for 3-6 weeks. Although some of the control group responded to the pseudo-stimulation, as expected (placebo effect), the patients receiving tDCS had a significantly better improvement in symptoms.

There is also such a thing as transcranial alternating current stimulation—tACS. This form of stimulation hasn't been well studied because it doesn't seem to be as potent as DC and the effects are more complex. But it's been used in a variety of techniques such as the old mind machines and, more effectively, in nerve stimulation. A more thorough investigation of the stimulus parameters may yield even better results in the future.

Thus far tDCS has had some success in clinical trials compared to the placebo effect, but researchers have to be careful. There can be some tingling associated with stimulation, which means some patients may be able to tell the difference between real stimulation and fakery. This isn't generally a concern for TMS, which doesn't use electrodes. But it's not a big problem in tDCS either. And like TMS, pseudo-stimulation with tDCS doesn't put researchers in an ethical quandary as is the case with DBS, which requires a craniotomy.

Why does tDCS seem to be lagging its TMS rival? If tDCS is as effective as the early trials suggest, you might think it would have been developed first rather than the more expensive TMS. However, it is this very simplicity and lower cost that has been causing at least some of the delay. As a treatment, tDCS may or may not be as effective as TMS, but since it's so cheap there are no expensive devices to sell. This means there isn't any profit incentive. Medical device companies aren't going to lay out big bucks to fund the kind of clinical trials needed for FDA approval of something that won't provide a significant return on their investment. That might seem unjust but it's a sound business policy. Would *you* invest your money in something that involved a lot of risk and had almost no potential for gain? If tDCS is effective it will eventually gain acceptance, though it might take a while.

\* \* \*

Stepping back for a moment to look at the big picture, we can ask how the noninvasive techniques described in the present chapter compare with invasive techniques such as DBS. Since DBS involves an invasive procedure to insert the electrodes, you'd think it would be at a huge disadvantage. And so it is, from the standpoint of medical risk and patients' psychology. But there's more to it than that.

DBS is much more precise and controllable. Those tiny electrode tips can deliver electricity to any small region the physician wants to stimulate. And with the sophisticated imaging and computer-controlled guidance systems, placement can be extremely accurate. With DBS you can deliver the payload exactly where it's needed. Noninvasive techniques can't do that. With TMS and tDCS you're stimulating a huge number of neurons encompassing a broad area of territory, which may also be home to neurons or passing axons that you'd very much like to avoid if you could. But with an imprecise technique you can't avoid them. It's like operating with blunt instruments rather than with razor-sharp scalpels.

Depth is another issue. In DBS you can insert electrodes anywhere you want, no matter how deep. True, you're going to pass through brain tissue and cause a little damage, but with thin electrodes this isn't a big problem. In contrast, the TMS device I described above only works to a depth of a centimeter (0.4 inches) or so.

This might change soon, however. An Israeli company called Brainsway has developed and patented a coil that they claim can reach greater depths of brain tissue without increasing the intensity of the fields to a dangerous level. Some researchers are calling this technique deep TMS, or dTMS. I've seen some reports on clinical trials with this technique, but at this point I'm a little skeptical. Although these reports describe successful results, they appear to do nothing but replicate the results of conventional

TMS. If it doesn't add anything, then unless it's more cost effective I don't know how useful it can be. Another problem is that I would think it could increase the risk of seizure—if you're stimulating more tissue, you're more likely to set off an avalanche of activity.

When depth and precision issues are considered, an invasive technique such as DBS doesn't look so bad. For movement and obsessive-compulsive disorders, I don't see it being replaced by noninvasive techniques any time soon.

How do the modern noninvasive techniques stack up against the old electrotherapy era? Even after all these years we still don't know what's going on in the brain when we stimulate it. We don't even know how the brain normally works, much less what happens when you excite or suppress it in some form or fashion. Our understanding of the mechanisms of brain stimulation is based on guesses.

But they're better guesses than ever before. Although our theoretical knowledge of the human brain remains unimpressive, we've developed excellent tools with which to judge the effectiveness of brain stimulation, and we have a much better handle on the placebo effect. Overall, I'd say the quackery quotient is considerably diminished from what it was in the 19th and early 20th centuries.

The future holds even more promise. And danger, perhaps.

# 10
# The Near Future: Enhancement, Augmentation, and Mind Nudging

If you can repair an ailing mind with electricity—or at least relieve the symptoms—is there any way to enhance a healthy mind? Maybe you'd like to increase your powers of perception. Who wouldn't want to be smarter? Or perhaps you want to boost your memory. Do important things sometimes slip your mind? Happens to all of us. Even if your brain is running smoothly and nothing is wrong, sometimes you forget someone's name, or a date, or another crucial bit of information that causes you to make a dreadful mistake. What if you could retain everything you learn? And what if you could accelerate the speed at which you learn?

This all sounds nice but rather fanciful. Except for fictional detectives such as Sherlock Holmes and Hercule Poirot, the human brain is far from perfect. It didn't evolve to memorize lengthy strings of random digits or solve complicated mathematical equations or absorb infinite amounts of data effortlessly. It evolved to control movement and to collect and retain enough information to stay alive, at least until its owner has co-authored like-minded progeny. Humans can perform astonishing feats of memory or calculation with the aid of devices such as pen and paper or a computer, but on our own we're quite limited. Short of a major overhaul of neural systems, it seems unlikely that you can get the brain to do much better than what it's capable of doing now.

It's not that the brain is awful. I've heard professors call the brain a "kludge," meaning that it's cobbled together from bits and pieces that don't fit together very well—a motley assortment of evolutionary hand-me-downs.[1] I think that's a silly argument, made by people who spend far too much time in their ivory towers mourning what they regard as the ignorance of us common folk. What the brain does, it does very well. And it does what it needs to keep us alive. The list of things humans can do is impressive: we can recognize faces, convey information by using a rich language, discover or invent practical solutions to problems, create novel objects, and imagine what may happen in the future. Our performance in most of these tasks exceeds even the fastest computer, and probably will continue to do so for some time to come. We're the most successful multi-cellular organism in the history of the planet. (Whether we last for long is another question.)

And yet, every once in a rare while a person is born with a brain capable of performing computer-like feats. On some occasions a person may also acquire an exceptional ability after experiencing some kind of trauma. These specially gifted people are often called savants, from a French word meaning to know. Some of them can factor large numbers in their head, or memorize tomes and recite long passages, or replay a complex piece of music after hearing it only once.[2] It's true, however, that savants generally display certain peculiarities and have problems with even the most basic social interactions, which is reflected in the old and entirely un-politically correct term for these people: idiot savants. It would seem that a reconfiguration of neural systems drastic enough to endow prodigious feats of computation or memory has its price. Which, I think, bolsters the point I made above—the brain is finely tuned for what it does.

But it goes to show you what might be possible. Maybe there's some way of tweaking the brain so that you can learn the skills of a savant without disrupting any other function. I doubt it, but only because it seems too good to be true, not because I can point to any specific physiological reason or theory. I can only say that most of the brain's parts appear to be so interconnected that it's unlikely you can make severe changes in isolation. But our understanding of the brain isn't sufficient to preclude the possibility of such things ever happening, even if they seem miraculous. (To me, the human brain has always seemed to be something of a miracle even without such enhancement.) And if there's even a remote chance of developing advanced skills without paying a heavy price, perhaps the judicious application of electrical stimulation within some set of parameters can prod our performance in the right direction. Maybe it can even sculpt neural systems and networks to produce these skills without compromising anything else.

Or perhaps we should set our sights a little lower. If you can't redesign the whole thing or reconfigure major parts of it without messing up something else, maybe you can slightly enhance performance with a tweak here or there. Fine tune the fine-tuning. And then, as our knowledge of the brain grows, we might become capable of making ever increasing improvements.

This chapter focuses on what I believe are the most likely developments in the near future. How will the techniques and results of brain stimulation evolve in the coming years? Will brain stimulation ever fulfill or exceed the hopes of pioneers such as Delgado?

There are also dangers, as Delgado well knew. Some of these dangers may also eventually come to pass.

Will brain stimulation one day become as addictive as, say, heroin? The rat experiments of James Olds and his colleagues suggest a frightening possibility. Picture those rats as they press the bar in their cage and self-stimulate themselves to exhaustion, even to the exclusion of food or sex. Could it happen to us? Science fiction writers have capitalized on such horrors with stories about "wireheads." What begins, perhaps, as a bit of fun and an automatic way of feeling pleasure could turn into a ravenous compulsion, as these unfortunate people become mindless zombies whose only goal is that next hit of electric current. I've already expressed my skeptical views on this in chapters 5 and 6, but if we going to speculate about *any* possible future, it's worth another look.

We should also consider stimulation that could enhance experience and increase performance without changing our underlying neural networks at all. This would require additional circuitry in the form of electronics. If we could plug our brains into devices such as computers, we could have access to their data storage and control their blindingly fast floating point operations. But first we need to build an interface or bridge between brains and machines that would permit two-way communication. The brain activity of the users would provide instructions (picked up with recording electrodes and interpreted by the computer), and the results would be encoded and directed back into the brain with electrical stimulation. Does this sound like science fiction? You might be surprised what is already being done these days.

The notion of electronic communication between brains and machines makes some people uneasy. What if the instructions began flowing the other way, from machine to brain? Requests may eventually become orders. Will we be able to resist? Delgado gained fame in the 1960s when he stopped a charging bull with remotely controlled electrical stimulation of its brain. Influencing human behavior with electrical stimulation is also certainly possible, as indicated by stimulation therapies for mood disorders. But how far can it go? Armies of electrically controlled minions?

Today's technology puts a limit on what can be accomplished at present. None of the speculative scenarios I've hinted at are doable right now, and the physiology and anatomy of the brain might be of a nature that makes these things infeasible for millennia, possibly forever. But if the brain is amenable to such manipulation, the technology of the not-too-distant future may find a way. That's why we should always be interested in new technological developments. And peek around the corner to see what might be forthcoming.

I've had to be selective in my choice of topics in this chapter, even more so than in the rest of the book. There is no end to the titillating stories and intriguing questions and fascinating speculations. But I had to make the book a *finite* size. For each idea presented, countless others had to be tearfully neglected.

* * *

In a science fiction story written by Philip K. Dick, Douglas Quail is a clerk who yearns to visit the colony on Mars, but he can't afford it. So he does the next best thing: he visits Rekal,

Incorporated, which promises to implant in his mind memories of a visit to Mars. He will not have gone to Mars but he would "remember" a trip, and receive various mementos of the "journey." Quail's brain work doesn't go as planned and leads to a strange revelation concerning Quail's real identity—a common theme in PKD's fiction—but the idea of creating memories is an interesting one. The story, called "We Can Remember It for You Wholesale," was originally published in 1966, and was the basis for the Arnold Schwarzenegger film *Total Recall.*

Although far-fetched, it's not impossible to imagine a process of sculpting new memories by pinpoint stimulation of specific pairs of neurons. Presently accepted theories of long-term memory formation involve some sort of strengthening or weakening of synapses. (Short-term memories, such as remembering a phone number long enough to punch it in, rely on other processes, such as reverberating brain activity.) The synaptic change represents the long-term memory—since synapses underlie communication, any adjustment in synapses alters the way the brain processes information. This is also the way you learn new skills. Researchers have found a number of ways synapses can change, many of which involve correlated activity in the neuron sending the message (the presynaptic neuron) and the recipient (postsynaptic). So in theory all you have to do to create memories is stimulate pairs of synaptically coupled cells.

Simple. But as of now we don't have tools that come anywhere close to the precision required to do this. And, as I've often written in this book, I can't believe the brain is really that simple anyway.

Although as of now we can't implant memories, we might be able to facilitate the process by which they are stored or retrieved. Scientists have already found that stimulating certain areas of the brain can enhance some types of learning and memory.

For example, Felipe Fregni, Alvaro Pascual-Leone and their colleagues published a study in 2005 that tested the effects of transcranial direct current stimulation (tDCS) on working memory.[3] Working memory is the neuroscientist's name for the process of keeping items such as phone numbers in mind until you're done with them. Fregni gave 15 subjects a task that involved remembering a letter they had been shown earlier, and tested their performance with and without tDCS. (The control group received sham or fake stimulation.) The part of the brain the scientists focused on is a familiar one—a region of the left frontal cortex called dorsolateral prefrontal cortex (DLPFC) that we have met before (see chapter 9, note 24). When the anode was placed over this area—this means the stimulation was generally excitatory—the subjects significantly outperformed those who received sham stimulation. The researchers also tested the effect of putting the cathode over the DLPFC, and discovered no significant enhancement. Anodal stimulation of the primary motor cortex also failed to deliver any benefits. In 2008 Suk Hooh Ohn of the Sungkyunkwan University School of Medicine in South Korea and his colleagues reported similar findings.[4]

Studies like these are fascinating but they don't tend to generate a lot of excitement outside of the scientific community. It's not like we're creating eidetic memory—another name for photographic memory, an exact recall of what one has just seen—instead, in these experiments the subjects were merely a little better at remembering whether they'd seen a certain letter in the recent past. Not earth-shattering stuff, but it does prove that noninvasive electrical stimulation of the brain can affect the kind of mental processing that we use every day.

If brain stimulation can slightly enhance some forms of memory, can it adjust the way people process certain types of information? If so, brain stimulation can not only augment skills, it can change perception. This is important because perceptions are crucial determinants of behavior. We would be closing in on a scenario that Delgado hinted at in his book *Physical Control of the Mind: Toward a Psychocivilized Society.*

Fregni, Pascual-Leone, and their colleagues published a study in 2008 that examined the effect of brain stimulation on a type of perception that a lot of people would like to influence—food craving.[5] Provocatively titled "Transcranial direct current stimulation of the prefrontal cortex modulates the desire for specific foods," the paper reported the results of experimental tDCS of the DLPFC in volunteers (21 of the 23 participants who began the study were women). The researchers tested how tDCS would affect the volunteers' thoughts about food and their eating behavior as they sat at a food-laden table, and after watching a video about delicious foods that typically invokes hunger. Since the study took place in Sao Paulo, the delicacies were Brazilian: *beijinho* (coconut candy), *brigadeiro* (homemade chocolate), *pacoca* (peanut candy), and other treats. Testing included the completion of questionnaires evaluating cravings as well as the number of calories ingested.

As is common with tDCS studies, the researchers split the experimental group into two, one of which received stimulation with the anode placed over the left DLPFC and the other with the cathode. (The other electrode was placed on the same site in the other hemisphere—the right side.) The control group received pseudo-stimulation. There was a clear difference in cravings as reported by the questionnaire and rating scale: subjects with the cathode on the left had less craving than the other groups, though the reduction was only 17.9 percent. The researchers also reported that "caloric ingestion" after both types of electrical stimulation was "significantly lower" than the control group.

It seems to me this study is a mixed bag and difficult to interpret. There appears to be only slight differences in the two experimental groups, which I would have thought would be more pronounced if the excitation/inhibition effects of switching the cathode and anode are important.

You might be wondering if the stimulation had a general effect on the volunteers' mood, which may have been the real cause of the drop in the desire to eat. But the researchers surveyed the volunteers' mood and reported no change. Still, I can't help wondering if the stimulation wasn't making some of them a little bit nauseous. The researchers, however, offered several other hypotheses: stimulation affected the "reward" circuits of the brain so that the delicious food lost some of its appeal, or stimulation resulted in an increase in the ability to resist temptation.

For whatever reason, this study didn't get much attention. But the next study I'll discuss got a ton of attention—which shows that modern electrotherapy is alive and well, particularly when it involves the notion that we can not only influence behavior with brain stimulation but we can drastically transform it as well.

Allan Snyder at the University of Sydney in Australia and his colleagues performed this study, which was published in 2003 and involves the "virtual lesion" effect of TMS, as discussed in chapter 9.[6] Removing a malfunctioning section of brain tissue makes sense in the treatment of certain disorders, but it can also help when that part of the brain is doing something to suppress desirable activity elsewhere in the brain. Although as I explained in the previous chapter there is some debate about exactly what repetitive TMS does to the brain, Snyder assumed that the low-frequency rTMS he employed effectively turned off the cortex at the site of application, which was a region of the left hemisphere around the border of the frontal and temporal lobe (the fronto-temporal lobe).

Snyder and his colleagues tested 11 right-handed men (all college students). Handedness is important because it reflects in many cases the hemispheric differences in the brain—right-handed people have a greater tendency to use their left hemisphere for language (see note 2 of chapter 4). Snyder's thinking on this issue reflects the standard and popular oversimplification of hemispheric specialization: left brain is for logical analysis, right brain is for creativity. I believe there is some degree of truth to this but—and here I go again—the brain is more complicated than that.

So the idea was to inhibit the logic-oriented hemisphere, thereby releasing the creative hemisphere from the fetters of cold, hard reasoning. Snyder and his colleagues wanted to create a savant out of a "normal" person with the aid of low-frequency rTMS.

After a 15-minute stimulation, the volunteers performed several drawing and proofreading tasks. A control group performed the same tasks but got sham stimulation.

In one of the drawing exercises, the volunteers were asked to draw a dog or horse from their own imagination; in another task, the volunteers reproduced a female face they had been shown earlier. An impartial committee judged the results. Four of the volunteers who had undergone TMS were judged to have changed their drawing style, becoming more complex and slightly more realistic. (After examining the samples published in the paper I can see a slight difference but it's really not much—a few extra lines here or there, or a more rounded curve.) Only one participant was judged to have changed after sham stimulation, but it wasn't consistently observed in all of his drawings.

The proofreading test involving catching duplicated word errors. The idea here is that people tend to miss errors because they read what is meant rather than what is there (even trained copyeditors occasionally fail to spot a mistake). Savants, according to Snyder, lack "our propensity to impose meaning and concept" and as result they could be better at catching errors. Two of the volunteers improved their proofreading skills after stimulation. These two were also in the group whose style of drawing had changed.

Why did so few respond? Snyder and his colleagues blamed TMS and its variability: "If the method of stimulation could be made more precise, all participants might have been influenced equally," he wrote in the report.

Snyder has received widespread media coverage on his work at the Centre for the Mind at the University of Sydney. The list of media outlets is impressive: *New York Times*, *The Times* (London), *Nature*, ABC News, CNN, and so on. In my opinion it's *far* overblown, yet another example of either journalistic credulity or the need to sell a story. The notion of producing savants by disrupting one of the brain's hemispheres is pure charlatanism. New Age snake oil. You're not going to unlock creative genius by the application of a pulsed magnetic field in a rather large part of the cortex. None of the participants in Snyder's study emerged as a gifted artist—the changes were marginal at best—and that's not due to the vicissitudes of TMS, it's because the idea is flawed with oversimplifications and trendy but false assumptions.

Despite my rant, I'd actually like to see more research on noninvasive stimulation methods. I loath the silly claims that people sometimes make and the foolish headlines written about them, but as part of a larger and sustained program with more realistic goals, this research can probe the limits of what is achievable with these devices.

Even with additional research, though, I predict we're not going to be using these devices in the near future to do much more than treat the symptoms of a certain set of psychiatric and neurological disorders. In other words, I doubt these devices will ever leave the clinic. This is particularly true for TMS, since TMS machines that are safe and easy to operate cost thousands of dollars. But transcranial electrical stimulators are cheap, and if they ever proved to be effective at enhancing our brains, they could become household appliances.

What I can imagine most of all is not the benefits but the headaches. And I'm not just talking about the users. How would the manufacturer protect itself from lawsuits after some less than brilliant users unwittingly gave themselves or someone else a seizure? If you limit the current to a level low enough to be almost completely safe for everybody, the machine would be nothing other than the placebo devices being sold on the market today.

I think other technologies that I'll describe later have more promise.

* * *

Brain-machine communication is a common prop in science fiction. The hero's brain is linked to a computer or an airplane or some other kind of machine which he controls with his thoughts. He can request information from a digitally stored encyclopedia or escape the bad guys with a slick maneuver in his instantly responsive mind-controlled flier.

Those gadgets probably won't hit the market for quite some time to come, but they are at least theoretically possible. And, I think, likely—inevitable, even, unless the present trends of our increasingly digitized and connected world somehow reverse course. Whether these devices will require invasive procedures depends on future technological developments. How cool it would be to slip on a noninvasive skull cap and have instant access to news and databases while your thoughts guide the electronic cook that's making your breakfast.

A two-way communication link would need a device to record or transmit your brain activity and another channel for information to flow back to you. The fastest way to get information into the brain is to use electrical stimulation—if you can somehow encode the information into signals the brain can understand and then deliver them to the appropriate neurons. Otherwise you're stuck with much slower sensory processes. For example, your request for information from the encyclopedia can be made directly through your brain-machine interface but the requested information might be presented on a standard computer screen. And feedback from the device you're operating with your mind can come from the usual sources—you make sure it's doing what you want by watching it and listening to it. In these scenarios you can talk to the machine but it doesn't talk back, so you have to use old-fashioned methods to get information from the machine.

A lot of work has been done on this kind of one-way interface that interprets your thoughts. It's often called a brain-machine interface or brain-computer interface, or sometimes it's called a link instead of an interface. Many people think of it as similar to a software interface, such as the operating system you use to control your computer. Operating systems generally employ graphical user interfaces these days, with the interaction between user and computer governed by icons and mouse clicking rather than inputs of text from a keyboard. In a brain machine interface, the interaction between user and machine comes straight from the user's brain activity. It would seem a natural progression of computational technology.

But science fictional scenarios aren't what motivate most researchers who work in the various fields of biological science. In most cases the goals are clinical. Brain machine interfaces would be exceptionally useful for patients who are paralyzed, such as those with spinal cord injuries, or patients with diseases such as amyotrophic lateral sclerosis[7] (ALS). Many of these patients are forced to rely on assistance for even the most basic tasks. A device that could enable them to do a lot more things for themselves would have huge benefits. Patients could regain mobility by being to able to operate a wheelchair, they could manipulate mechanical arms to grab and move things, and they could more easily use communication equipment. Not only would this technology reduce medical costs, it would also provide a great boost for the patients' psychology, letting them achieve some measure of independence. Independence is a wonderful thing, a fact known all too well by people who have lost it.

This kind of one-way brain machine interface is a fascinating area of research, but since it doesn't involve stimulation, it's off topic as far as this book is concerned. I'll confine myself to just a few comments on it.

In the 1990s, John Chapin, Miguel Nicolelis and their colleagues began using large arrays of electrodes inserted into the brain to record neural activity as animals performed tasks such as pressing a lever. Recording the action potentials of many neurons at the same time had been pioneered several decades earlier (by many researchers, including my thesis advisor, George Gerstein), but Chapin and Nicolelis had the specific goal of learning how to decode the outputs of these neurons and using the data to control a device.[8] In a paper published in 1999, they described an experiment in which they recorded neurons from a rat's motor cortex as it pressed a lever.[9] The lever moved a robotic arm which delivered water to the thirsty animals. The researchers then used a mathematical function on the recorded neural firing activity to extract a signal by which they could interpret the rat's intentions. In the next step of the experiment, the researchers set up the robotic arm to be driven by these signals rather the rat pressing the lever. In other words, the rat controlled the robotic arm with its brain rather than with lever presses. Of course the rats didn't know the switch had been made—at least not at first. But the researchers discovered that when the robotic arm was in the "neurorobotic" mode, the rats eventually learned that they didn't have to move the lever any more. Think, and ye shall receive.

A similar but even more complex experiment was conducted and published in 2000 by Nicolelis and his colleagues.[10] In this experiment the researchers recorded from a large number of neurons in several areas of a monkey's cortex, including motor cortex and neighboring regions. They again managed to extract signals that represented the monkey's intention to make a specific arm movement, and used these signals to operate a robotic arm.

With these and many other experiments, researchers have inched their way toward developing interfaces to help disabled patients. But the kind of signal extraction in the rat and monkey experiments of Chapin and Nicolelis requires precision recording of neurons—and this means electrodes into the brain, not resting against the skull. Electrodes pasted on the skull can record the electricity activity of the brain that seeps through the cranium, but the signals are weak and come from a broad area of the under-lying cortex. As I mentioned earlier, skull-recorded signals—which are generally called EEGs—contain a mishmash of electrical activity from a large number of neurons carrying out a large number of different functions. It's extremely difficult to dis-entangle this potpourri. Some would say it's impossible. An additional problem with using the EEG for this purpose is that it's contaminated with signals from nearby muscles, which are also electrically active. EEGs are great for observing large-scale brain activity and checking for seizures, but not so great for detailed analysis of neural networks. Some researchers have tried to use EEGs to drive brain machine interfaces and have met with modest success. But there are limits to EEG precision. To go further, we need an invasive technique, which once again leads to craniotomy issues. It's a problem that has yet to be resolved.

Research on the other direction of information flow—from machine to brain—has also been making some progress. I've already described many such devices in the chapter on prostheses. Cochlear implants and artificial retinas convert rather simple bits of information—frequencies of sound or pulses of light—into electrical signals that the brain can interpret. This constitutes information flow from machine to brain. Some people accordingly label these prostheses as brain machine interfaces, and so they are, in a limited, one-way direction.

When you put both recording and stimulation together you get an interface capable of a bidirectional flow of information. A real synergism of brain and machine. Many people use the term *cyborg* to describe a combination of machines and biological materials or systems.[11] I don't particularly care for this word because science fiction writers and movie-makers have adopted it to such an extent that it now carries all kinds of futuristic connotations. The same goes for the term *bionics*.

I suppose these terms adequately convey the notion, though in a somewhat exaggerated sense. But you don't need to build a brain-machine combination with such a permanent weld that neither is functional on its own. Start small. Designing a bidirectional interface is hard enough.

Once again, Nicolelis and his colleagues are pioneers in this endeavor. They have performed an extension of their earlier experiment with monkeys to include stimulation in addition to recording. As described in a 2009 article, the researchers implanted arrays of electrodes into the brain of a monkey and taught him to use a joystick to position a cursor on a screen.[12] One of the tasks required the monkey to move the cursor left or right depending on a cue. If the cue was present, the monkey moved the cursor in one direction, otherwise he moved it in the other. At first the cue was delivered by the vibration of the joystick. After the monkey became proficient at the task, the researchers began using the recorded brain activity to control the cursor instead of the joystick, similar to the experiment described earlier. But they also started cueing with stimulation rather than the joystick. When the monkey's brain was stimulated he moved the cursor (with his brain activity) in one direction, otherwise he moved it the opposite way.

It's important to understand the limitations of this experiment.

The stimulation didn't provide a coded set of instructions, the monkey was merely sensing the presence of the stimulation and acting accordingly. But it's also important to understand that the stimulating currents were low—it was microstimulation. It's not like the monkey got a big jolt in the head as a cue. This is evidenced by the failure of another monkey in the experiment to master the task with a stimulation cue—he never learned that the stimulation was providing the cue. But his stimulating electrodes were placed in a different site in the brain than the other monkey. The stimulation in the successful monkey was delivered to the primary somatosensory cortex, while the monkey that failed to learn was stimulated in a region of parietal cortex a little farther back from the primary somatosensory cortex.

The problem of conveying information or instructions within some kind of pattern of stimulation is hugely more difficult. In the experiment with the monkey, the stimulation was like a switch—on/off, yes/no—one bit of information. To provide additional information requires much greater complexity.

And there's a serious obstacle. As I've mentioned several times in this book, individuals vary. Our brains don't work alike. They operate on basically the same principles but the details vary depending on training, experience, and genetic influences on anatomy, wiring, and biochemistry. The way your brain processes information is not the same as the way mine does. Therefore, to transmit information to a given brain you've got to discover its code. I don't know exactly how much variability there is among individuals—this is an open question. The human imaging experiments with fMRI and PET might give you the idea that our brains are quite uniform, but those scans and the pretty pictures made from them are deceptive. They're based on averages. Not only does the activity vary from individual to individual, it also varies from trial to trial, even when the person is doing exactly the same thing. Scientists who do imaging experiments average out all this variability using sophisticated statistics.[13] The human brain will not be easy to read.

But the great advantage of brains is that they are so trainable and adaptable. Maybe it would be a lot easier if you could just teach the person—and his or her brain—the machine's code. When I stimulate this neuron, it means such and such, and that neuron, it means something else.

Is this possible? How sensitive are people to what's happening in their brain?

Questions like that always make me uneasy. I start wondering what difference there is between a person—his or her identity, consciousness, and so forth—and the brain. Obviously consciousness doesn't require every single neuron in the whole brain or you would be in trouble because neurons die every day. So are "you" just a subset of neurons, some sort of special ensemble that oversees the other, serf-like workers?

I could ramble on this topic for pages but that's for another time and another book. Let me just mention an interesting finding of P.R. Kennedy and R.A.E. Bakay.[14] In the course of treating a female patient in the latter stages of ALS—when the patient was almost completely immobile—the physicians attempted to communicate with her by way of an electrode implanted in her brain (specifically, the motor cortex). Although she couldn't move, perhaps she could learn to vary the activity of her neurons and use that as a way to signal yes or no in response to questions. The patient was provided visual feedback: the researchers used a screen to display the output of the electrode as it recorded the firing activity of a small number of neurons. The patient could also hear the output because the signals were also sent to a speaker (an action potential makes a popping noise when the electrical signal is converted to sound). With the help of this feedback she gradually became able to increase or decrease the firing rates of these neurons with (the rest of) her mind. Such variations can be used to communicate with people or operate a device. This was a case of exquisite control over one's brain.

Perhaps the ultimate interface is virtual reality. In the present state of the technology, though, it has nothing to do with stimulating the brain. Virtual reality is a catchphrase for more or less realistic simulations of various environments. It involves one or more sensory modalities—usually vision, as presented on a small screen as the user wears a helmet, for example—and massive amounts of computation to recreate, as much as possible, certain aspects of some experience, whether it's flying a plane or diving under water. In the high-end, expensive versions, the computers respond as users shift their gaze, allowing varying perspectives, as if you were actually present in the environment. In my opinion, few other technologies except possibly nanotechnology have received as much hype. But the simulations can be impressive. Even at the lower, cheaper end of the spectrum of virtual reality, such as computer games, you can find yourself drawn into the simulation. People call it immersion.

Science fiction authors have imagined futures in which virtual reality is much more than just putting on a helmet or inserting a glove. In these stories a computer stimulates the user's brain to recreate every little detail of the simulated environment. It's like you're actually there.

But there's a problem. It's not possible to do this with the kind of simple devices portrayed in these stories, where the interface is some easily removable skull cap—yeah, right—or an array of electrodes inserted into some small region of the brain. You can't transmit enough information to enough neurons with this kind of configuration.

There's another big problem with such futuristic scenarios: you can't fool the brain as easily as some science fiction writers think you can. It's going to be extremely difficult to make people actually believe they are *in* a simulated environment. I suspect users will always be able to tell the difference between the real thing and a simulation.

The problem with fooling the brain with a simulation is that the brain has a lot of back-up systems. If vision fails you, you've got hearing, or smell, or taste, or touch. You've also got senses you probably don't think of too often, such as proprioception—the ability to sense your body's position in space. All these channels of information get constantly updated and processed, whether you are conscious of it or not. If all's well then you're feeling good, but if not, you've got a bad feeling. (And sometimes more than that, as in seasickness.) Ivory-towered professors may regard the brain as some clunky contraption, but they're wrong—their philosophy is clunky, not the brain. The brain excels at what it does because it evolved over a long period, mostly in a world where getting fooled or confused meant becoming dinner for a saber-toothed tiger. Saber-toothed tigers are gone now, but the legacy of the danger they and other predators once presented left an indelible mark on the human brain.

To fool the brain into believing it's somewhere it's not, you're going to have to control all the inputs, and make them so realistic that the complex processing centers can't detect a hint of contradiction. I believe such a high level of technology and scientific knowledge will be a long time in coming.

\* \* \*

Although no virtual reality technician or computer specialist can control all of the incoming sensory data of a person's brain—not even a mad scientist can do that yet—partial control is possible. And this is enough to wield a certain amount of influence on what a person is thinking and how decisions are made. It's hard to dominate a brain but you can manipulate it if you are subtle enough.

I'm not just referring to marketing gimmicks and glib politicians. It's bad enough that people throughout history have been taken in by scams, seductive advertising, false prophets, deranged but charming manipulators, snake oil salesmen, and Democrats and Republicans. If you can prevent people from getting any information other than your propaganda, if you can limit a population's educational opportunities, and if you are clever enough to appeal to their prejudices (and we all have them—yes, that even includes *you*), people can be tamed and herded like sheep.

You can also manipulate behavior by a judicious application of brain stimulation. This is a documented fact, as I will explain below. But how much of a problem it could become in our society is debatable.

Mind control is an old topic. Falling under the spell of a witch or the dominion of a demon has been considered a dreadful possibility throughout history. Think of the witch trials in colonial Salem. In modern times, political ideology has taken the place of witches and demons. During the Korean War in the 1950s, for example, some American prisoners of war appeared to cooperate with the enemy and spoke out against the American government. Intelligence agencies in the United States feared the worst: the enemy had discovered some kind of potent "brainwashing" technique. There were several possibilities, including mind-altering drugs and hypnosis.

Adding fuel to the fire, Richard Condon's novel *The Manchurian Candidate* told the story of captured American soldiers who had been sent to a secret camp in the Chinese region of Manchuria and turned into brainwashed assassins, to be loosed on an unsuspecting United States. (The novel was published in 1959. The story was filmed in 1962, and remade in 2004.)

Fiction is one thing, but real-life stories also began appearing. During his trial for the murder of actress Sharon Tate and several others, a diminutive con-man and aspiring musician named Charles Manson appeared to exercise almost mystical control over young women in his "family." In another case, Patty Hearst, heiress to the Hearst newspaper fortune, was kidnapped by the Symbionese Liberation Army in 1974 and participated in armed robbery, after which she claimed in court to have done so against her will.

It was in this zeitgeist that some people improbably attributed the microwave radiation at the U.S. embassy in Moscow, described in chapter 9, as an attempt at manipulation of behavior.

Although the microwave radiation couldn't possibly have done much, some POWs and others who have been subjected to strict confinement might well be described as manipulated. But there are no secret thought rays or drugs. It's all about control over the victim's environment. Relentless indoctrination and prolonged deprivation are enough to instill a sense of helplessness on anyone. Everybody has a breaking point. At some juncture you give in because you no longer have the energy to resist. It's like getting into an argument with an utterly dogmatic person who refuses to admit he's wrong and summarily rejects all evidence that is contrary to his beliefs; unless you have an infinite amount of patience then you'll eventually reach a point where you say, "Yeah, whatever," and walk away. But when you're incarcerated you can't walk away.

The job of a manipulator is a lot easier when the person to be manipulated is cooperative and willing to believe. Manson took in a lot of young people, mostly women, who were troubled and unsure of themselves. He told them what to believe and restricted their access to anyone who might have contradicted him. And that's all there was to it. Add a little charm, lots of sensual pleasure from sex and perpetual intoxication from drugs, and there you go. No mystery there.

But what about the use of brain stimulation to expedite the process? Is this something we need to be concerned about in the future? Imagine the power of being able to control people without having to dominate their environment. Will advances in neuroscience and technology imperil our freedom?

I've tried to take a balanced approach in this book, although I haven't hidden my deep skepticism of some of the bolder claims people sometimes make. I'm also inclined to dismiss the notion that brain stimulation can achieve absolute control over minds. Even Delgado refused to accept this idea (although at times he seemed to hint otherwise). There won't be any zombie armies of brain-stimulated fanatics. The point I made with virtual reality applies here—a would-be dictator couldn't control all of the human brain's channels of information and its complicated processing networks. Which also goes back to the point I've made repeatedly (ad nauseam?): the brain isn't that simple.

Besides, the conventional way of producing zombie armies is easier. Indoctrinate and deprive. All you need is a group of impressionable people, and there are lots of them around. How else can you explain the huge number of fanatics in the world today? The United States certainly has its share too. Go to a political convention of either of the two main political parties and you'll find plenty of zombie minds.

But there's a caveat here, and it's more potent than the standard "you can't dismiss *anything* completely" argument. Stimulation-induced mind control isn't possible any time in the near future, and probably not ever, but you can sway minds, or at least modify decisions. Mind control, never; mind nudge, yes. The only question is the extent.

A number of experiments over the last few years have started to find some answers—and raised a considerable amount of controversy. In 2002 Sanjiv Talwar, John Chapin, and their colleagues made a big splash when they steered a rat by remotely stimulating its brain.[15]

The researchers operated on rats and implanted electrodes in two sites: the primary somatosensory cortex and the medial forebrain bundle. Recall that the somatosensory cortex contains a map of the body and receives inputs from nerves that detect touch and pressure. Talwar and his colleagues placed the electrodes in the part of the map that represents the whiskers, to which a disproportionate amount of territory in the map is devoted because whiskers provide important sensory information to these animals (similar to the large areas devoted to hands in a human's somatosensory cortex). The medial forebrain bundle should also be familiar if you remember chapter 5—this bundle of axons carries information between regions that are known to play a role in reinforcement and reward. Rats will self-stimulate with some of the highest response rates when the electrodes are implanted in their medial forebrain bundle. This being the case, experimenters have often trained rats to perform various behaviors by rewarding correct responses not with food or water but with stimulation of the medial forebrain bundle.

Talwar and his colleagues trained the rats to follow cues given by electrical stimulation of the whisker area of the somatosensory cortex. The stimulation was the electronic equivalent of reins— stimulating the whisker cortex of the left hemisphere signified a command for a left turn, and the right hemisphere for a right turn.

Stimulation of the medial forebrain bundle provided the reinforcement during training. The animal wore a receiver and microprocessor on its back—equipment Delgado referred to as "stimoceivers"—and the operator controlled the stimulator with transmissions directed by a laptop computer. The range was 500 meters (1640 feet). Trained rats could be guided through mazes and pipes as well as over rough terrain to reach a target destination. As Talwar wrote, "We have used this paradigm to develop a behavioural model in which an experimenter can guide distant animals in a way similar to that used to control 'intelligent' robots." In addition to the scientific aspects of the experiment, the researchers identified some practical applications, "such as search and rescue in areas of urban destruction and landmine detection." In other words, a rodent version of a Saint Bernard. But one with greater mobility, especially when compared to the awkward robots of today: "Combined with electronic sensing and navigation technology, a guided rat can be developed into an effective 'robot' that will possess several natural advantages over current mobile robots."

There's that word "robot" again.

This paper touched off a storm of media attention. Some journalists dubbed the rats "Roborats," perhaps making a reference to the Robocop series of movies, in which a dead policeman's brain is installed into a robot. Nicolelis added a "t" and called it "robotrat".[16] Videos of these rats in motion were included in the 2006 BBC documentary *Human Version 2.0*. Having watched the video, I have to say that the rats don't look very mobile. Instead of scampering around like normal rats, the "robotic" rats shown in this documentary lumber as if they were old and arthritic. Perhaps they were. But they've also undergone some major operations and are carrying around a lot of equipment on their backs, as well as on their heads.

I paid a lot of attention to this story for several reasons. For one thing, it reminded me so much of Delgado's earlier work, but nobody mentioned that. It didn't surprise me that journalists failed to remark on the similarity, but I would have expected neuroscientists to be familiar with this old work. Perhaps they considered it ancient history, which in biological research means a couple of decades or sometimes a lot less, depending on the field. Or perhaps most neuroscientists aren't interested in history, or subconsciously suppress their knowledge of the bad old days of the go-go era.

Another reason for my interest was that I knew Sanjiv Talwar personally. He and I earned our doctorates in the same lab at roughly the same time. I was delighted for his success; he had struggled with some difficult problems in his thesis research but he persevered, and he's a genuinely nice person. I ran into him a little while after the paper came out, and I jealously kidded him about his fame and all the interviews of him that had been published. To my surprise, he seemed uneasy about the whole thing. But I probably should have expected that—a lot of the media attention revolved around questions about the ethics and morals of using animals as "robots." Controversies have a way of unsettling nice people. It's no wonder he felt a little like Frankenstein.

One of the most interesting and potentially troubling aspects of this experiment is the kind of reinforcement that animals were given. It wasn't a tangible or substantive reward, nothing that satisfied a basic need, but instead consisted of electrical stimulation of the brain (in addition to and at another site from the "navigating" stimulation).

But these are just rats, right? Although rats are more like humans than we—and probably they—would care to admit, there's no denying our greater neural complexity. This greater complexity means that rudimentary approaches are unlikely to work. But in my opinion it also means that subtler approaches can succeed and, thanks to the complexity of the human brain and thought processes, these subtle influences might also be extremely difficult for the manipulated victim to detect.

Primates such as monkeys have neural networks much more similar to ours than rodents. In 1990 William Newsome and his colleagues reported that microstimulation of a certain part of the cerebral cortex of rhesus monkeys can affect their perceptual judgments.[17] The part of the brain the researchers stimulated is called the middle temporal visual area or MT, and is known to process visual information related to motion.[18] Like most cortical areas, MT consists of an orderly arrangement or map, but in the case of MT the firing rates of the neurons represent the motion of visual stimuli. Each small area of MT contains neurons that increase their activity when they detect something in the visual field that is moving in a certain direction. This small area of neurons extends all the way vertically through the cortex, forming a column of neurons that respond in a similar way. Different columns respond to movements in different directions, and the whole of MT contains columns that encompass all directions.

The monkey watched a screen that showed a bunch of spots that tended to move in one of two directions, and the monkey was rewarded when it shifted its gaze to a target on the screen that corresponded to the movement of the spots. (Researchers could follow the gaze of the monkey by tracking its saccades or eye movements.) To make this task a little harder, the spots moved randomly except for a tendency to drift in one direction or the other. Using thin electrodes inserted through an opening in the monkey's skull, the researchers recorded neurons from MT and determined their preferred direction—when the dots moved in this direction the neurons elevated their activity. MT neurons respond vigorously to these stimuli and clearly play a role in the processing of this information.

So what happens if you stimulate the monkey's brain while it's performing this task? You can obviously get the monkey to make a mistake by disrupting the processing of visual infor-mation—give the cortex a good jolt and you'll get one confused monkey. But that's not what Newsome did in this experiment. Instead, he stimulated a tiny area in MT—a column of neurons whose activity represented motion in a certain direction. If you stimulate this column, will it bias the monkey's judgment? Newsome and his colleagues discovered that this is what hap-pened. If you stimulate a column that represents motion in, say, the left direction, the monkey tends to judge the motion of the spots as being left-ward more often than on trials when this column isn't stimulated. The monkey doesn't respond randomly or not at all, which would have been the case if the stimulation had invoked confusion; instead, the stimulation biased the animal's perceptual judgment.

Nobody knows what the monkeys were seeing or thinking. If he missed on a certain trial because of the stimulation, I wonder if he would have said—if he was a human—"Wow, I could have sworn those things were moving in that direction."

Hossein Esteky and his colleagues performed a similar experiment and reported it in 2006.[19] The researchers stimulated neurons in a region of the lower temporal cortex called inferior temporal cortex (IT). IT is known to be an important area for the visual recognition of objects, and some of the neurons in this region are selective for faces—they respond vigorously when a face appears in the visual field, but not to other stimuli. Esteky trained monkeys to judge whether stimuli represented faces or not. The stimuli were noisy or distorted, so this was not always an easy task. The researchers discovered that stimulation of face-selective neurons biased the monkey to judge a stimulus as a face.

Do these experiments have any relevance for humans? I think they most certainly do. Like humans, monkeys process information in their brains along complicated pathways. A little stimulation at some point along the way can easily sway a perceptual judgment. A primate mind can definitely be nudged.

What makes it more eerie is that you probably couldn't detect it. A little bit of stimulation—microstimulation—generally isn't going to be felt. In other words, it doesn't quite make it into consciousness. It can influence your decisions but you don't feel its stimulating effects directly. And if you don't know the stimulation occurred, you'll probably accept the decision you made as your own. Which means you might have to get creative in how you explain "your" decision.

Delgado and other researchers found instances of this in their human patients. Recall the patient described in chapter 6 who came up with all kinds of reasonable explanations for why he turned his head. In reality, he turned his head because his brain had been stimulated, but he didn't perceive the stimulation, so he searched his mind for alternative explanations and hit upon the ones that seemed the most likely.

A similar phenomenon was reported in 1998.[20] Neurosurgeons were operating on the brain of a 16-year-old girl in the course of treating her seizures when they discovered they could invoke laugher by stimulating a certain part of her frontal cortex. (A local anesthetic was used so the patient was conscious.) The operation was part of a procedure in which the physicians were testing the effect of stimulation while the patient performed various tasks such as reading text or naming objects. Although the physicians knew the patient's laughter corresponded to stimulation, she did not, and her "laughter was attributed to the particular object seen during naming ('the horse is funny'), to the particular content of a paragraph during reading, or to persons present in the room while the patient performed a finger apposition task ('you guys are just so funny...standing around')."

We humans are great confabulators. Sometimes I think we never really know our own minds. Have you ever made a decision and then later, when you had to defend it, found yourself rationalizing it by citing arguments that didn't consciously occur to you when you made the decision? I've done that. You say to yourself, "Well, this argument makes sense so that must have been what I was thinking."

If mind nudging is difficult to detect, how much of a problem could it become in a future society where brain stimulation is much more prevalent than it is today? I suppose that largely depends on how prevalent it becomes and how precise it can get. These subtle nudges I've described required tiny currents to be directed at precise spots in the cortex. That's not going to happen these days without doing a craniotomy and inserting thin wires into the brain. And *that*, of course, is detectable.

Or is it? Perhaps technology will progress to the point where precision stimulation can be accomplished in sneaky ways. I'll have more to say on that in the next chapter.

But first, here's a scary thought. Perhaps technology has *already* progressed to this point and we don't know it.

Such ideas are the basis for movies such as *The Matrix* and its sequels, where machines control human beings by simulating "reality." But it isn't just Hollywood or science fiction authors who have considered this possibility. Philosophers often use the phrase "brain in a vat" to debate the nature of reality. What if you are just a brain floating in a nutrient-dense fluid? Could you tell the difference between being in a vat and having a body? Perhaps some alien or mad scientist or intelligent machine is stimulating our brains and creating our "reality." To do this they would need to have complete control over our environment and a full understanding of how brains work. Sounds far-fetched, and I've argued elsewhere in the book that this is for all practical intents and purposes impossible, but...you can never be certain. And there is nothing I know of that would theoretically preclude this from ever happening.

Maybe we're not even brains in a vat. Suppose we're just part of some computer simulation and we don't know it. It's not hard to imagine a future civilization so advanced that they could do this. Maybe we don't really exist except as software agents inside some gigantic computer-like device. Nick Bostrom is a philosopher who has argued that this is a distinct possibility, even a likely one. If the simulation is good enough we'd never know, would we? You could imagine a variety of purposes that these simulations could serve, from recreating evolution to just playing games.

It makes me wonder if this highly advanced, futuristic civilization that is supposedly conducting these simulations might not be a simulation itself. Suppose *they* are also actually simulations. Who is running the program? What is reality? Is anybody or anything really real? After all, if our reality is actually a simulation, then the real world, if such a thing even exists, doesn't have to be anything like the world as we imagine it here in our little simulation. The laws of physics and the dictates of logic that rule our simulated reality may be entirely fictional. Not only could the real world be different than we think it is, it might be something we would consider impossible.

And so you have absolutely no facts on which you can base any conclusions. Which is one of the reasons why I tend to regard this line of speculation as pointless.

# 11
## The Not-So-Near Future: Changing Minds

Brain stimulation will have an increasingly significant impact on society if the techniques continue to improve, as I believe they will. Although invasive brain stimulation probably won't ever become commonplace, TMS and tDCS may drastically change how people in the future will live. These tools will probably always be too crude for the type of perceptual influence that I've called mind nudging, but that doesn't mean they won't ever be useful to modify more general forms of behavior.

One possible application that would have a huge effect is the rehabilitation of criminals. Maybe we can reduce criminal behavior by enhancing the activity of neural networks responsible for morals and ethics and civic duty, or by inhibiting neural networks that make people more inclined to cheat and steal. This isn't possible today because our knowledge of the brain is so limited, but it might be possible at some time in the near future.

Could we all agree that this would be a good thing? I take it as a given that most respectable citizens would prefer to live in a world with far less criminal violence. But I suppose some people might still resist the use of brain stimulation for this purpose; perhaps they fear potential abuses by a totalitarian government, or perhaps they're more bothered about deeper philosophical issues concerning free will. Anthony Burgess explored this issue in his 1962 novel *A Clockwork Orange* (later made into a movie). In my opinion, population growth and crowding will one day necessitate some kind of behavioral modification unless we want to live in a world full of violence. Eventually we'll have to make some hard choices. Brain stimulation may be one of those hard choices.

A related topic is lie detection. The capability of the human brain to deceive and tell a lie is quite interesting, and in my view neuroscientists have yet to explore this subject as much as it deserves. Some people don't seem to be good liars, while other people appear to be highly proficient at it. Body language can sometimes give the guilty party away—things like blushing, smiling, fidgeting, failing to make eye contact—but not always, and it varies from person to person. Even professionals such as police officers who often deal with people of questionable integrity can have trouble spotting a liar.

Technology has tried to come to the rescue. In general, it's been a failure. In the 1920s came the polygraph—the "lie detector"—a machine that measures anxiety by monitoring blood pressure, skin conductance (a measure of perspiration), pulse, and respiration. The idea is that a person will be more anxious when he's lying. Problems with this concept abound: some people aren't more anxious when they lie, and some people are anxious nearly all the time, especially when someone is accusing them of lying. Research has shown that the polygraph machine is fallible, and it is not generally accepted as evidence in U.S. courts unless both parties agree in advance to its use.

Then came the idea that you can use brain activity to detect a lie. This makes sense—the brain is the source of all behavior, so couldn't you find evidence of deception by analyzing brain rhythms? Unless you're self-delusional or severely intoxicated, you know when you're lying, which means somewhere buried in all of that brain tissue there are at least a few neurons that know what's really going on. Lawrence Farwell has developed a system he calls brain fingerprinting which is based on the EEG. So far his success has been limited. In my estimation, Farwell has a good idea but EEGs are too crude for the job.

How about expensive imaging devices such as fMRI? A number of researchers have worked on this idea, and there are companies that offer lie detection services based on this technology. Once again, I feel like this is a good idea but we lack the right tools; imaging devices that are available today lack the precision and resolution to do a proper job.

Now comes brain stimulation. Research into the use of brain stimulation in lie detection has just begun, but the idea is simple and clear: use stimulation to activate neural networks associated with truth-telling and/or inhibit those involved with deception. Stimulation would be done noninvasively with TMS or tDCS.

Unfortunately, this might prove to be difficult. There's no such thing as a "truth center" or "deception center" in the brain. It would be nice if there was, because then you could use an imager, or perhaps even EEGs, to observe activity in these regions. When cortical area X lights up in the scanner, you're lying. How simple. Pity the brain doesn't work that way. People have looked long and hard for such tell-tale regions of the brain and have never found them.

High-level decisions such as whether to tell a whopper and the best way to go about it are the domain of complicated neural networks scattered throughout the brain. The frontal cortex plays a major role, as revealed in imaging experiments—the frontal cortex is often activated in these situations, though not in any manner that would let you detect when someone is lying (at least not with the tools presently available). Other regions are also involved. Planning and carrying out a deception requires a large number of operations, including the suppression of tell-tale body language, memorization of the fictitious story, and the restructuring of associated knowledge to accommodate the lie.

I have my doubts about how successful brain stimulation can be as a lie detector considering the limits of the present state of technology. But you don't know until you try.

A general thesis about lying is that it requires more brain power—more time and resources—than telling the truth. If that's true, it's reasonable to assume that liars have a longer reaction time and greater neural activation. If a person you are interviewing pauses before answering an important question, you're probably more likely to suspect a lie. A person who is telling the truth doesn't need all that time. (Unless, of course, the person is merely deciding the best way to say the truth, in which case the cause of the delay is in finding the right words rather than formulating a lie. Which goes to show that snap judgments based on a single data point can lead you astray.)

If stimulation could scramble the neural networks involved in this process, the deceiver might have a harder time of it. A report published in 2008 by Alberto Priori and his colleagues showed that tDCS increased reaction time in deceivers (but not truth-tellers) when the anode was placed over the dorsolateral pre-frontal cortex bilaterally.[1] (The stimulating electrode was split into two so that both hemispheres received the same stimulation. The other terminal in the circuit was placed on the shoulder.) Sham stimulation as well as cathodal (inhibitory) stimulation of the dorsolateral prefrontal cortex failed to have an effect.

A paper published a few years later reported a startling finding that stimulation of another region of the prefrontal cortex actually increases a person's facility to tell a lie.[2] Ahmed Karim, Niels Birbaumer and their colleagues used tDCS to stimulate the anterior prefrontal cortex. As the name suggests, this region of cortex is situated in front of the dorsolateral area, though it is still part of prefrontal territory. Imaging studies have linked the anterior prefrontal cortex to deception. Subjects in this experiment participated in a mock crime where they had the opportunity to steal some money and were then interrogated. The interrogators asked a series of yes/no questions and the subjects had the leeway to decide for themselves whether to be honest or tell a lie. During this interrogation the researchers stimulated the subject's brain in one of three ways: cathodal tDCS of the anterior prefrontal cortex (which tends to produce inhibition), anodal stimulation (excitation) of the same region, and sham stimulation. The researchers measured skin conductance (as with the polygraph) and reaction times during the interrogation. They also scored the subjects for their skill in telling lies and tested their feelings with questionnaires. As incentive to do a good job, subjects were allowed to keep their booty if they escaped detection.

Inhibition of the anterior prefrontal cortex resulted in a significant reduction in reaction times of deceivers (but not for truth-tellers) compared to sham stimulation. Strange as it seems, inhibition of this area made the subjects *faster* at making up lies. Not only that, but they were scored as having better lying skills, their feelings of guilt at telling lies were reduced, and their skin conductance was lower, which is generally considered to mean that they were less anxious. Anodal stimulation of the anterior prefrontal cortex did not produce these effects.

It's conceivable that cathodal tDCS of this region might simply be accelerating *any* kind of demanding task, and so the effect would be general instead of specific to deception. If this is true, you could claim that the reason why deceivers had an easier time in this experiment was because lying requires more effort. This explanation would be less exciting than one which proposes that the anterior prefrontal cortex is somehow specifically involved in the moral decision to tell a lie. The researchers explored this question by testing the subject on another demanding cognitive task that didn't require deception. If cathodal tDCS of the anterior prefrontal cortex affected performance during this task, the effect would clearly be more general. However, the stimulation had no effect, which means that as far as the researchers could determine, the results of their experiment were specific to the act of deception.

What does this mean? The researchers propose that the anterior prefrontal cortex is involved in moral judgments. Inhibition of this region may inactivate neural networks that suppress antisocial behavior. As the researchers put it, "deceiving another person in order to obtain personal profit seems to create a moral conflict, and if a person is relieved from this moral conflict, he/she might be able to deceive unhinderedly with faster RT [reaction time], less feelings of guilt and less sympathetic arousal as demonstrated here."

But why didn't excitation of the anterior prefrontal cortex produce the opposite result rather than having no effect? (There was a tendency for excitation to decrease lying skills and produce greater guilt but these data did not reach statistical significance.[3]) The researchers suggest that these networks may already be operating at or near maximum morality, so increasing the activation won't result in any further gains. I don't find this explanation convincing. It's obvious that other networks are involved, which was already apparent from imaging studies that clearly show that a host of sites become active during deception as well as any other complex behavior. As I've mentioned before, there is no lying center or morality center or anything of a similar nature.

In another experiment, Inga Karton and Talis Bachmann used rTMS to test the effects of brain stimulation on spontaneous lying.[4] The subjects were stimulated in the left or right dorsolateral prefrontal cortex while viewing colored discs; the task of the subjects was to name the color. During this experiment the subjects had the latitude to tell the truth or a lie. The researchers found a greater propensity to lie when the subjects received stimulation in the left dorsolateral prefrontal cortex (as compared to stimulation of the left parietal cortex), and a reduction in lying when they received stimulation in the right dorsolateral prefrontal cortex (compared with right parietal cortex stimulation).

These experiments show that stimulation can influence a person's propensity to lie as well as their skill in doing so. But this field of research is just getting under way. There are all sorts of ways people can tell a lie. It can be carefully planned with an intricate web of deceit or it can be a spontaneous decision with little forethought. A person who feels wronged might take a different emotional and cognitive approach to deception than a person who is seeking gain or leverage rather than revenge. There can be instances where lying is part of the job (secret agents and politicians, for example). Lies can come in different categories, from whoppers to little white lies ("yes, dear, that dress is lovely"). All we know so far is that stimulation can in some instances subtly alter the process in *some* circumstances, admittedly artificial. But better methods may be discovered at any time.

Suppose that in the near future we find a way of using brain stimulation to inhibit lies. Should it be applied at, say, criminal trials? Should it be voluntary or mandatory?

Although such speculations might seem pointless until we've successfully developed the technology, I believe it's better to begin such discussions beforehand. And if we decide a technology isn't going to be applied, there's no reason to spend any time or money to develop it.

Some people might argue that the 5th amendment to the U.S. Constitution would prohibit forcing a suspect in this country to undergo brain stimulation. The 5th amendment grants certain rights to defendants, including the right not to be "compelled in any criminal case to be a witness against himself." In court jargon, taking the fifth means refusing to answer questions. Does brain stimulation violate this right? I don't think so, any more than compelling a suspect to provide DNA samples or undergo blood, urine, or breath tests, all of which have been ruled perfectly legal. You can't be forced to volunteer information, but that doesn't mean law enforcement agencies can't get it out of you by other means. But I'm not a legal scholar. Brain stimulation may be regarded as different from other medical procedures because the brain is the seat of consciousness. Issues might also be raised with the 4th amendment, which protects citizens from "unreasonable searches and seizures", though I'm of the view that 4th amendment rights are routinely violated these days anyway.[5]

Strong arguments can be made for and against brain stimulation in this and other applications. As for brain stimulation in criminal cases, I'm for it. It isn't so much that I approve of brain stimulation or I'm all for using it in every possible situation, but in this particular instance I'm willing to overlook some of the thornier arguments against its use. In my opinion the American justice system is weak and distorted with unscrupulous lawyers, intellectually challenged juries, and convoluted jurisprudence, so any help at arriving at the truth would be a tremendous blessing.

But there is always the potential for abuse. The same is true with any application of brain stimulation. The extent of the danger, though, is far from clear. Earlier in the book I explained why I believe brain stimulation will never be likely to enable mind control or create a class of people helplessly addicted to electricity ("wireheads"). Yet I have to admit that the possibility of mind nudging worries me. How far can it go?

Not to put too fine a point on it—and I certainly don't want to foster paranoia—but lots of emotions as well as cognitive thought processes could be affected. For instance: whenever you feel good, do you always know why? Let's use our imagination here and suppose that someone could influence your mood *without your knowledge*. I don't mean the ability to make you deliriously happy or suicidally depressed—I don't see that happening any time soon, and probably not ever—nor is it likely that you could be forced to do something against your will. But what if a sneaky bit of brain stimulation could perk you up a bit? Or bring you down a tad, perhaps. Rewards and punishments, even slight ones, can go a long way to shape behavior. Psychology is full of examples clearly demonstrating the role of reinforcement in governing human behavior. If you repeatedly smile and show approval whenever someone incidentally makes a certain gesture, that gesture will tend to be repeated, even though the person won't generally realize why he or she is doing it.

Not only is it possible to affect moods with brain stimulation, it has already been done and is continuing to be done clinically these days. But it requires techniques that can't be performed without your knowledge. Nobody is going to perform a craniotomy on you or paste electrodes to your scalp without your permission.[6] But what if somebody develops a method that doesn't require invasive surgery or a trip to the clinic?

* * *

Warning: this last section of the book is speculative.

This doesn't mean I'm letting my imagination run wild. My speculations will be constrained by what neuroscientists have discovered about the brain and what I believe is a reasonable extrapolation of our present knowledge and technology. I will be guided by my beliefs about how our society will evolve, though I don't claim to have any special ability to predict the future.

I don't believe that we've gone as far as we can go in the applications of electrical brain stimulation, although I've repeatedly stated my skepticism in this book of bold futuristic claims. Such claims are in my opinion based on the assumption that the brain is an incredibly simple device. Some people seem to believe that all we have to do is find the right button to push, and voila, we get mind control, addictive stimulation, miraculous cures, flawless prostheses. But no such buttons exist. Something as complex as the brain requires complex techniques and instruments to influence and manipulate it.

What is the future course of brain stimulation? How might its influence and number of applications grow? As I see it, the primary consideration is the need to avoid a craniotomy yet somehow stimulate specific neural networks and systems.

Avoiding a craniotomy means transporting the source of stimulation across an intact skull. We already know how to do this, and several chapters of this book have described techniques using fields and radiation. The problem is specificity. Noninvasive techniques use fields, or radiation if the source of stimulation must be far away, but we don't have enough control over the effects to do anything that requires precision.

Weak fields or radiation won't accomplish anything, and if made too strong, they can destroy tissue. Moderation is critical. But even if you manage to find the ideal intensity, that doesn't solve the problem. It's nearly impossible to hit just one spot. Electromagnetic energy can be focused with lenses or it can be created as a coherent beam (as in a laser), but it will nonselectively activate neurons as it travels through brain tissue. It's not the same as inserting a slender electrode that is insulated everywhere except at the tip. With the electrode you get much more control than you can achieve with stimulations based on fields.

Perhaps there might be a way to get around this. Imagine aiming a lot of low-intensity beams at a single spot in the brain. The individual beams are too weak to stimulate neurons so there is no undesired activation as they travel through tissue. At the meeting point, however, the intensity of the combined beam may be sufficient for stimulation. The stimulation target is the confluence of the beams.

This technique would be complicated and require not only precise anatomical maps but also the willingness and ability of the person to maintain his or her head in just the right spot. It's not a great option, particularly if you want to be sneaky about when and where you stimulate someone.[7]

Even if it worked there would still be a specificity problem. I haven't presented much neuroanatomy in this book and I'm not going to start now, but what is important is that any given area of brain tissue, no matter how small, contains a wide variety of neurons and/or axons. Some of the neurons are connected and form part of a network or circuit having a common goal or function, but other neurons in the area may be involved in exactly the opposite function—inhibitory neurons, for example, are often located nearby to tamp down activity. Passing axons may be from neurons that are involved in an entirely different set of operations. Even neurons that are part of the same network can play much different roles. Unless you can achieve precision to the level of a tiny fraction of an inch, the fields will still activate an undifferentiated mass of neurons and/or axons without regard to their function or connections. Precision will likely never be possible with any form of radiation. As a result, the use of noninvasive techniques will therefore always resemble a sledgehammer rather than the targeted precision instrument we require.

The solution to this problem is conceptually easy. You use weak fields that won't activate anything they're not supposed to, and amplify the strength of the field near or in those neurons you desire to stimulate.

How can this be accomplished? I suspect we'll need to put something in the brain that can be triggered by a weak form of electromagnetic energy. This means that the energy should cause these substances to change form or do something that activates them and sets into motion certain biological processes that will result in an electrically active neuron body, axon, synapse, or whatever we wish to stimulate.

Substances that react to electric or magnetic fields or radiation are common. Let's confine the discussion to radiation since it has unlimited distance and is easy to generate and employ in a stealthy manner. Generically, molecules that respond in some manner to radiation are called photoreactive, and they are plentiful in nature. Chlorophyll, for example, absorbs light and is a critical part of the process by which plants extract energy from sunlight. Opsins in photoreceptors of the retina change form when they absorb light, which is the first step in the retina's transduction of light into electrical signals. Other examples include molecules with chemical bonds that are susceptible to breaking when exposed to radiation.[8]

Finding the right photoreactive molecule would allow us to target small areas. If we could use some kind of substance that is involved with or related to specific neurons having a specific function, we could pick and choose exactly what we want to stimulate. Such a procedure is definitely feasible, since neurons do often have biochemical specificities related to their function. For example, neurons employ various neurotransmitters such as dopamine or acetylcholine and associated molecules, such as enzymes that synthesize these neurotransmitters and other enzymes that destroy excess amounts (neurons don't want these powerfully active neurotransmitters floating around for very long). Peptides (tiny proteins) and other molecules can also be specific to a neuron's function.

To make this work, though, we need some form of radiation that penetrates the skull and at least partway into brain tissue yet doesn't do any harm. The key issue is frequency. The energy of radiation is directly proportional to its frequency—the higher the frequency, the greater the energy. This is not the same thing as amplitude or strength of the beam. Using a weak beam of X-rays or other high-energy radiation will not produce the same effect as a strong beam of low-energy radio waves. High-energy radiation penetrates biological tissue but also has the capacity to break molecules apart or strip electric charges from them, a process called ionization. Ionizing radiation can cause serious damage to tissues and is also difficult to control.

We need some form of low-frequency radiation. Radio waves and other extremely low-frequency radiation penetrate tissue with ease, but this is because they have little energy and engage in few interactions. Microwaves are a possibility but their energy is still low. Light would be perfect but since we don't have transparent heads, it wouldn't work too well. There is, however, a "window" at a range of frequencies close to visible light. Radiation known as near infrared can penetrate the skull.[9] It's not ideal because bones do scatter some of it, and it doesn't penetrate watery tissues very well so it mostly reflects from the surface of the brain, but it can reach the cortex and maybe a little below.

Researchers have taken advantage of this technique to image the brain in ways similar to functional MRI. The first step is to shine a beam of near infrared radiation at the subject's skull. The spectrum of the radiation that is scattered from brain tissue depends on the amount of blood that is present and its oxygenation. By measuring the scattered radiation scientists can measure blood flow in cortical areas, which is basically what a metabolic imaging device such as fMRI does, and is useful in the same way as fMRI—blood flow is associated with neural activity. The use of near infrared in this manner is called near infrared spectroscopy, often abbreviated NIRS in the scientific literature. As a tool to image brain activity it has some splendid advantages over fMRI: it is vastly less expensive and does not require the subject to sit still for a long period of time (which is extremely important when you're trying to image small children).

Unless future research discovers a more effective range of the electromagnetic spectrum, near infrared is likely to be used in the kind of subtle brain stimulation I'm describing here. It's easy to generate—everything emits infrared radiation, particularly warm objects, although unless they are red-hot they don't emit much in the near infrared portion of the spectrum. Perhaps there might be problems with controlling the amount a person receives and ensuring the right dose, but I don't see any insurmountable difficulty.

The radiation delivers the energy to the special molecule that is in the cortex of the brain, waiting for activation. But which molecule should be used?

Researchers have already found a number of interesting

photoreactive molecules that directly affect a neuron's excitability. Many of these molecules originally come from microbes. Some single-celled organisms have dramatic responses to light, and often contain kinds of opsins that are linked to important activities in the cell. These activities can include the flow of ions. A particular kind of opsin called channelrhodopsin has been discovered that activates an ion channel in response to light. This was an interesting discovery on its own, but what made it even more important is that geneticists have long since developed techniques to transfer genes from one organism to another. Genes that are normally found in microbes can be inserted into mammalian cells. When you do this with one of the various types of channelrhodopsin, you get a cell that changes its electrical activity when light of a certain frequency is shined on it. Put that in a neuron and you've got a brain cell whose electrical activity you can optically control.

This amazing experimental technique is presently being applied to the study of brains and neural networks in vitro and in vivo. Scientists can insert the gene into neurons with the aid of various viruses that have been engineered for the purpose of delivering genes and gene regulatory sequences (similar to the still developing form of treatment known as gene therapy), or develop a line of transgenic animals that express the gene in their brains. Activation is generally done through a craniotomy, using fiber optics to deliver a carefully measured quantity of light to a tight space. The beauty of it is that you can record the electrical activity without interfering with the optical controls. With this approach researchers can switch a neuron on and off while monitoring other cells in the neighborhood. A general name for experiments that use light-activated molecules in this manner is optogenetics, a combination of inserting the proper gene and exercising optic control. Karl Deisseroth of Stanford is a pioneer in the application of optogenetics to the study of mammalian nervous systems.[10]

Optogenetics is a productive avenue for scientific research but isn't quite what I had in mind for the futuristic brain stimulator.

Transgenic humans are a bit far over the horizon, at least in my estimation, and besides, optogenetic techniques generally require a craniotomy to expose the brain to the engineered viruses and the fiber optic cable. But it's the right concept—you add something to the brain and then activate it with electromagnetic radiation. The difference is that the brain stimulator I'm considering would need a frequency of radiation that can pass through the skull. And the use of genetic tinkering to create the something that's added to the brain seems a bit far off in the future to me, though DNA technology is proceeding at a rate that's quite astonishing (some would say alarming). It's entirely possible that gene transfers might be involved in prepping a person for brain stimulation in the future.

Surprisingly, some frequencies of infrared radiation appear capable of stimulating neurons without any neural modification at all. Anita Mahadevan-Jansen and her colleagues at Vanderbilt University have reported that pulsed infrared radiation can generate action potentials in peripheral nerves without injuring them. The radiation did this all on its own; the pulses apparently create a bit of heat that transiently opens up channels in the membrane. Recently Mahadevan-Jansen and her team have used infrared radiation to stimulate somatosensory cortex in the rat.[11] But the researchers had to expose the animal's cortex in the procedure.

The generic term for stimulating physiological tissue with light is photostimulation. From the few brief examples I've given here, you can tell that photostimulation is being increasingly applied to neuroscience research. But the futuristic device I've thinking about goes well beyond the present technological capacity of today's laboratories. Let's call it Brain Activation by Stealthy Stimulation (BASS). The patient—or subject, or victim— swallows a pill or gets an injection that makes certain neurons more excitable when irradiated by electromagnetic energy of a frequency that at least partially penetrates the skull.

Important reiterating note: I'm not advocating BASS. I'm just playing what if. It's interesting to think about this kind of technology because in my estimation it's feasible, which means that one day we might have to deal with it whether we like it or not.

Delivery of the photoreactive molecule presents several difficulties. There are the usual issues of dosage, metabolism, and side effects. In general, drugs enter the body, circulate everywhere (and interact with non-targeted organs or systems, causing side effects), and then are eventually eliminated, often after bio-chemical degradation that produces still more side effects. Even if the subject is a willing participant, drugs can be tricky and fail to work as intended. An additional problem arises when the target is the brain. The blood-brain barrier filters the blood supply to the brain and precludes large molecules from entering. This protects the brain from junk floating around in the blood that might disrupt the delicate neural biochemistry. It also frustrates neuro-pharmacologists to no end.

There may be a way of getting around the blood-brain barrier. It's sneaky—as sneaky as a virus. Certain viruses like herpes simplex and rabies appear to get into the brain via the nerves. By taking the back-door they avoid getting filtered out by the blood-brain barrier. Perhaps it may be possible to use these viruses as a delivery system, once the disease-causing genes are ripped out. This is an idea that has plenty of precedent—gutted viruses have long been used as delivery systems for such procedures as gene therapy.

Let's assume for the moment that the issues with drug de-livery can be solved. What sort of behavior or emotion or thought process can we influence when we expose the subject to the radiation?

Mood will probably be easy to influence. Some people call serotonin the "happy chemical" because it is a neurotransmitter involved in carrying messages between neurons that regulate mood.[12] Although it's not that simple, these and certain other molecules play an important role in a person's disposition. BASS might affect these neural systems by regulating the quantity or availability of these chemicals, or it could activate small molecules that bind proteins involved in synthesis, transportation, release, or uptake. Any of these options or some combination thereof could have a strong impact on mood.

But many of these systems operate deep in the brain, perhaps out of reach of the radiation. Unless the radiation BASS uses can somehow manage to attain more depth than near infrared, it will be limited to cerebral cortex.

In my opinion that's not a serious limitation. The cortex is a critical component in most of our complex thought processes. Judgments, perceptions, and decisions are the domain of the cortex. These processes impact your whole life, including mood.

I'll give you one particular example. Once when I was in college I lost my wallet. I came home from the library and when I reached into my back pocket to put the wallet on the dresser, I was horrified to discover it wasn't there. It occurred to me that I probably left it lying in the library where I had been feeding dollar bills to the copy machines. Morbid thoughts of financial ruin suddenly tormented me; all of my pitifully small savings was accessible with the bank card, and the credit card could put me in debt. I would be broke and homeless.

I rushed to the table where I had deposited my car keys, hoping to race to the library and retrieve the wallet before it was stolen. It was then that I saw my wallet lying next to the keys. For some inexplicable reason I had broken with tradition and absent-mindedly set the wallet with the keys.

Relief flooded over me. I felt a tremendous lift in mood. Prior to this I had been in a middling sort of mood, but now I was elated. Such was my relief that the good feeling lasted the rest of the day.

But later that night I began thinking how odd it was that I should be feeling so good. I was not in any way better off—I merely found something I thought I'd lost. Since there was no real change in my circumstances, my mood should have returned to what it had been before. Instead it overshot the mark by quite a bit.

The brain activity and biochemistry involved in regulating mood is complicated, and I'm sure that some physiological "overshoot" had occurred as the system attempted to return to equilibrium. Equally obvious, the cortex had been involved. The nature of the threat to my welfare had required imagination—the problems lay in the future, and were only recognizable after going through a chain of logical reasoning (lost wallet, possible theft, access to bank account and credit, no money left, how will I be able to pay the rent and buy food in the future?). Although describing subcortical structures as "primitive" is an over-simplification, they are not primarily involved in logical trains of thought involving the future. The threat wasn't something that neural networks deep in the brain can easily interpret—the threat wasn't big and hairy with long sharp teeth—but instead came from my frontal cortex ruminating over a dire future.

The cortex can strongly influence mood. If you can stimulate it then you can adjust how a person feels, and with that power gives you the ability to shape behavior.

If BASS could stimulate structures deep in the brain, it could do a lot more. It might even be deadly. What if BASS could stop ion channels or other basic electrophysiological processes in brainstem neurons that govern critical functions such as respiration? What if a similar technique could be used on ion channels in the cardiac cells that generate the heart beat? You could stop a heart instantaneously. Assassination would be easy.

For now this is pure fiction. My point is that it may be possible.

But the difficulties of stealthily getting the photoreactive molecules into the victim's brain correctly and in the right dose might be too challenging. Transmitting the radiation might prove too much of a problem as well. Powerful fields may be required, generated by equipment that could be too difficult to hide, especially if it needs to be close to the victim.

A system with a willing participant would be easier to design. This kind of brain stimulation could be used as a treatment for neurological or psychiatric disorders, or as a means of enhancing neural function. The stimulation would entail the same precision and selection as BASS—an external stimulator activates some kind of chemical incorporated into specific neurons—except the procedure could be performed openly and in an obvious manner because it was requested. Let's call this Brain Activation by Manifest Stimulation (BAMS).

What could it do? The possibilities are legion. Although I don't believe you could control a person's thoughts or create a permanently happy state—the rest of the brain would know something funky is going on—you could accelerate certain processes such as learning or memorization, you could heighten sensitivity to certain perceptions, you could influence cycles such as sleeping and waking, and much else. The changes might last only during the stimulation, or the stimulation could trigger physiological mechanisms that endure for a long time afterward.

If you can activate multiple targets in an intricate way, I suspect you could even make fundamental changes in a person, provided the person is willing to go along with it. Perhaps the stimulation might be able to alter an entire personality. As a person who sometimes experiences radical shifts in thinking, I've always wondered if such changes could be triggered at will by some complicated stimulator. BAMS has a chance of succeeding. Whether this is desirable or not is another question.

Suppose you have a job where you need to be a stern taskmaster. Let's say you're able to do this, but you have a tendency to bring your work home with you, and your spouse and the kids are less than enthused. Your family life suffers. I believe a lot of us have problems of this sort, where we need to adjust our behavior to suit different requirements but we have great trouble doing so. Although the ideal solution may lie outside of technology (find another job, or try to adapt the circumstances to your personality rather than vice versa), one way of dealing with it could be BAMS. Tough guy at work, nice guy at home. All you have to do is swallow a pill and place your head briefly against a radiator.

I realize that such technologies open up a broad range of questions regarding ethics, philosophy, and the law. Some of you may be horrified at the very thought of this kind of brain stimulation. Those of you who feel that way might be even more horrified to learn that when some people hear about this kind of stuff they think, "Cool!"

As for me, I'm sort of in the middle. Sometimes I think that's not a good place to be; I'm reminded of the old adage that the only things in the middle of the road are lane markers and flat possums. At other times I think there are not enough of those of us who don't behave like ranting and raving ideologues. I guess that's why I find myself in the middle so often.

The philosophical issues may be the hardest of these issues to address, and they're also the most personal, which makes them much less likely to be resolved by logical debate. Who you are is in part who you think you are and how you see the rest of the world. The idea of brain stimulation to such a degree as I've hypothesized in this section may completely upset your beliefs of personhood. On the other hand, you might regard it as just another tool to help you succeed, no more than a calculator or a power saw.

Ethics, legal and otherwise, is probably more amenable to a rational discourse on the potential fallout from a new technology such as BAMS. But I've mostly avoided the subject in this book. A professor in my graduate school once half-jokingly said that the university was required to offer a certain course in ethics but nobody in his department was qualified to teach it. I feel the same way. What this book has been about is the science and technology of brain stimulation, but I'm well aware that there is more to life than this. I just don't think about the other stuff very often. There are other people to do that, and that's fine with me.

Will BAMS ever get implemented, even if it is technologically feasible? I'm not willing to offer any specific predictions. I think BAMS or something like it is doable within a few decades if researchers invest enough time in it (and are able to procure enough grant money to do so). Whether people will embrace or reject this technology is up to that unfailingly chaotic entity we call society.

# Notes

Chapter 1

1: Brad isn't his real name. In this and other descriptions of actual patients in the book I have changed some of the minor details to protect privacy.

2: Electroencephalography is the process of monitoring electrical brain activity with electrodes pasted to the subject's scalp. A record of this activity is called an electroencephalogram. The abbreviation EEG can refer to the process or to the recording.

3: PET can be used to image the activity of any organ or tissue in the body. The procedure involves radioactive atoms that release a positron as they decay. (A positron is a particle of antimatter—it is the antiparticle of the electron.) The radioactive atoms are incorporated into a biologically active compound such as glucose and injected into the body. These radioactive compounds are used by the body in a normal manner—the radioactivity has no effect on the biochemical activity of the compound. For example, radioactive glucose is taken up by cells and broken down for energy. Meanwhile, the radioactive atoms decay, releasing positrons that combine with electrons to produce particles of light known as photons. These photons have a specific frequency and can be detected with sensitive instruments. PET scanners track these photons back to the site where they originated, and the number of these photons gives a measure of the activity of these sites. For instance, neurons use a lot of glucose when they are active, which means that an injection of radioactive glucose will collect in these cells and emit a lot of photons. Thus a PET scan detects metabolic activity in the brain, which tends to correspond with electrical activity (neurons that are active need more energy than cells that aren't).

4: A neural network refers to a group of neurons that communicate via synapses and process information. Many people like to think of it as a circuit, with neurons as electrically active elements. Note that the neurons in a network might all be close neighbors or they may be scattered throughout the brain—the only thing that binds them are the synaptic connections and their specific function (which is related to the specific kind of information they process, such as sensory information or motor commands). Engineers and computer scientists have adopted the term *neural network* to describe certain types of computer software or electronics that emulates the information processing capacity of neural networks. This technology, more properly called artificial neural networks, has been used in a variety of computational tools designed to mimic how the human brain processes information.

5: MRI is a technique that uses a strong magnetic field along with radio waves to image parts of the body. The magnetic field aligns protons in hydrogen, which interact with the radio waves to produce signals that a computer can use to visualize the structure of tissues. An MRI machine provides high-contrast images of soft tissues such as the brain, which often makes it more useful than X-ray devices. The MRI image is a snapshot. To visualize *activity* rather than just a static image, you need to take a rapid series of images. This is what fMRI does. With this technique, you can measure the oxygen levels in the blood supply of the brain, which is a measure of metabolic activity. Active neurons have a higher metabolism, so fMRI, like PET, is an indirect measure of neural activity.

6: A neuron's cell body is about 50 micrometers (0.002 inches). The cell body, or soma, contains the nucleus, which houses the chromosomes, and other structures important for the complex biochemical activities needed to maintain life. Most neurons also have extensive branches, or processes, that extend from the body. Dendrites are extensions that tend to be specialized to receive inputs from other neurons—many synapses are located at dendrites. An axon is an extremely thin extension of a neuron that carries the neuron's signal—which is a series of electrical pulses called action potentials—to other cells (often making synapses on another neuron's dendrites, but also sometimes on its soma). Dendrites can be thick or thin and form a dense layer around the soma or can extend some distance away, though usually not too far. (The word *dendrite* comes from a Greek word meaning tree.early anatomists who peered through microscopes evidently believed dendrites looked something like a tree extending from the soma.) Axons can travel for however long they need to go—all the way from one side of the head to another, or from the brain down to the spinal cord, or from the spinal cord all the way to a toe.

7: Alexander Volta. 1800. "On the Electricity excited by the mere Contact of conducting Substances of different kinds." *Philosophical Transactions of the Royal Society of London.*

8: Mary A.B. Brazier. 1958. "The evolution of concepts relating to the electrical activity of the nervous system: 1600 to 1800." *The Brain and Its Functions.*

## Chapter 2

1: James P. Morgan. 1982. "The First Reported Case of Electrical Stimulation of the Human Brain." *Journal of the History of Medicine.*

2: Roberts Bartholow. 1874. "Experimental Investigations into the Functions of the Human Brain." *The American Journal of the Medical Sciences.*

3: Orrin Devinsky. 1993. "Electrical and Magnetic Stimulation of the Central Nervous System: Historical Overview." In *Advances in Neurology,* volume 63: *Electrical and Magnetic Stimulation of the Brain and Spinal Cord,* edited by Orrin Devinsky, Aleksandar Berić, and Michael Dogali.

4: The cerebellum—the "little brain" that is tucked between the back of the cerebral hemispheres and the spinal cord—also has a cortex, called the cerebellar cortex. But in this book when I write "cortex" I'm referring to the cerebral cortex.

5: David Ferrier. 1876. *The Functions of the Brain.*

6: For example, an editorial published in *Nature* on 6 August 1874 was titled "Hitzig v. Ferrier."

7: Roberts Bartholow. 1874. "Experimental Investigations into the Functions of the Human Brain." *The American Journal of the Medical Sciences.*

8: Roberts Bartholow. 1874. "Experimental Investigations into the Functions of the Human Brain." *The American Journal of the Medical Sciences.*

9: James P. Morgan. 1982. "The First Reported Case of Electrical Stimulation of the Human Brain." *Journal of the History of Medicine.*

10: James W. Holland. 1904. "Memoir of Roberts Bartholow, M.D." *Transactions of the College of Physicians of Philadelphia.*

11: James W. Holland. 1904. "Memoir of Roberts Bartholow, M.D." *Transactions of the College of Physicians of Philadelphia.*

Chapter 3

1: You may have heard of neurogenesis, which refers to the addition of new neurons. This can happen in the adult human brain, but the process doesn't involve division of a mature neuron but rather divisions of stem cells which result in cells that evolve—biologists say "differentiate"—into neurons.

2: You may be wondering about the difference between a contusion and a concussion. Both are serious types of head trauma, but a contusion is localized—it shows up on brain scans as a bruised region. A concussion is a more diffuse injury, and the damage is more widespread.

3: Stephanie Pain wrote an article on this in the 16 September 2000 issue of *New Scientist*. I'd like to believe that at least some of these trepanning cases were medical. I realize society has changed over the ages, but making a hole in someone's head just for the sake of some sort of ritual couldn't have been all that popular.

4: Depending on the spot of stimulation, the experimenters probably would have invoked movements on both sides of the face, though primarily on the right side, contralateral to the site of stimulation, which was in the left part of the brain. Although it is true that each hemisphere of the brain controls the opposite side of the body, the face is slightly different in that some of its muscles receive contributions from both halves of the brain.

5: Michael Powell. 2006. "Sir Victor Horsley—an inspiration." *British Medical Journal*.

6: These days surgeons use local anesthetics.

7: For example, in 2002 the U.S. Senate considered a bill that would have made the use of birds, mice, and rats in animal research much more expensive. Senator Jesse Helms sponsored an amendment to the bill that nullified this potential burden on researchers. The Society for Neuroscience, of which I was a member at the time, sent an email to members asking us to urge our senators to vote for the Helms amendment. I thought it was amusing that a scientific organization whose members tend to be so liberal in their politics would champion legislation sponsored by such a staunch conservative as Jesse Helms.

Chapter 4

1: Penfield struck a note of teamwork with the title of his auto-biography, *No Man Alone: A Neurosurgeon's Life*, published in 1977.

2: In most people, the language center of the brain is found in the left hemisphere. This is true for almost all right-handed people and also for the majority of left-handed people, though the percentage is not as high as for those who are right-handed. In a few people language seems to be bilateral, or in other words, the responsibility of both hemispheres. There are lots of theories to explain these curious facts, but no one has yet come up with a convincing explanation. One way of determining which hemi-sphere controls language is the Wada test. Named after the neurologist who developed it, the test involves anesthetizing one of the cerebral hemispheres. This can be done by injecting anesthesia into the appropriate carotid artery. When one of the hemispheres is "asleep" the subject's language ability is tested. If the ability to use language has been greatly impaired, the anes-thetized hemisphere is presumably dominant for language skills.

3: Wilder Penfield and Phanor Perot. 1963. "The Brain's Record of Auditory and Visual Experience." *Brain*.

4: The area or areas Penfield worked on depended on where the patient seemed to be having trouble with seizures, which could be narrowed down before the operation by EEG or by the specific sensations or muscular contractions the patient experienced during seizures.

5: I got information on these patients' experiences from several sources, including the following two articles:

Lamar Roberts. 1961. "Activation and Interference of Cortical Functions." *Electrical Stimulation of the Brain*, edited by Daniel E. Sheer.

Wilder Penfield and Phanor Perot. 1963. "The Brain's Record of Auditory and Visual Experience." *Brain*.

6: Lamar Roberts. 1961. "Activation and Interference of Cortical Functions." *Electrical Stimulation of the Brain*, edited by Daniel E. Sheer.

7: Wilder Penfield. 1968. "Engrams in the Human Brain." *Proceedings of the Royal Society of Medicine*.

8: Electricity doesn't flow in biological tissue the same way it does in a wire, although scientists often model it that way. Electrical phenomena in the body obey the same rules as in other material, but the mechanisms are different than in circuits made of wires and batteries. Ions carry the current in biological tissue, but cell membranes form serious obstacles—and tissue is crowded with cells—so ions must take alternate routes. Membranes made of fatty molecules, as is the case with cellular membranes, form something like a brick wall as far as electric charges are concerned. Current can flow across membranes only through ion channels, but these channels may or may not be open depending on the state of the cell.

9: Wilder Penfield and Phanor Perot. 1963. "The Brain's Record of Auditory and Visual Experience." *Brain*.

10: Wilder Penfield and Phanor Perot. 1963. "The Brain's Record of Auditory and Visual Experience." *Brain*.

11: Wilder Penfield and Phanor Perot. 1963. "The Brain's Record of Auditory and Visual Experience." *Brain*.

Chapter 5

1: The surface of the living brain looks pink because of the blood vessels. Honestly, I've never really understood this "gray matter" designation; though I have seen tissue slices that appear grayish, they usually look more brown than gray. In a chemically preserved brain the white matter is really white (sort of), but the cortex and other cell-containing material definitely looks brown to me. But I suppose "brown matter" will never catch on—there would be no end to the scatological jokes.

2: The term *stereotaxic* comes from Greek words meaning solid and orientation—it's a method of finding your way around a solid body. An alternative spelling is stereotactic.

3: Every individual's brain is slightly different anatomically, so it's still not easy to hit the target, particularly if it's tiny, as it often is. But modern neurosurgeons have a wealth of sophisticated instruments such as computerized scanning machines and magnetic resonance imaging (MRI) to guide them, so it's not so much of a problem as far as human operations go these days.

4: This is not necessarily because neuroscientists have any reservations about using felines. Rats are much cheaper.

5: Today the lingua franca of biology is English, and working biologists from all countries need to know this language. The reason is that the United States spends far more money on biomedical research than any other country, and so most of the scientific papers in this field of research are written in English.

6: Nobel Lecture, 12 December 1949.

7: Konrad Akert. 1961. "Diencephalon." *Electrical Stimulation of the Brain*, edited by Daniel E. Sheer.

8: Olds later found out that the electrode was accidentally embedded in axons from the rhinencephalon—the "smell brain," which is often associated with regions of the brain involved in emotion.

9: James Olds. 1956. "Pleasure Centers in the Brain." *Scientific American.*

10: James Olds. 1956. "Pleasure Centers in the Brain." *Scientific American.*

11: Maya Pines. 1973. *The Brain Changers.*

Chapter 6

1: How times have changed. The success rate, for example, of grants to U.S. government funding bodies such as National Institutes of Health (NIH) averages about 20-30 percent. In 2011, the success rate for NIH was 18 percent, even more dismal than normal—the NIH funded less than one in five proposals submitted to them that year.

2: Video clips from this experiment can be seen in a variety of places. For example, the 2006 BBC documentary *Human Version 2.0* includes video from Delgado's experiment. The web site Youtube is another good place to look.

3: José M.R. Delgado. 1969. *Physical Control of the Mind: Toward a Psychocivilized Society.*

4: José M.R. Delgado. 1963. "Effect of Brain Stimulation on Task-Free Situations." *Electroencephalography and Clinical Neurophysiology.*

5: The central gray is also known as the periacqueductal gray.

("Gray" because it is gray matter—it contains cells.) It is located deep in the brain and surrounds a fluid-filled space called the cerebral aqueduct. Parts of this region are known to be involved in processing painful stimuli or formulating behavioral responses to such stimuli.

6: José M.R. Delgado. 1967. "Aggression and Defense under Cerebral Radio Control." *UCLA Forum in Medical Sciences, Number 7: Aggression and Defense.*

7: Other explanations are possible. Perhaps the lower-ranking monkeys were constantly testing the alpha male in subtle ways—not enough to earn a scolding but just to see if he was on guard. When alpha failed to meet the challenges, the other monkeys lost their fear of him.

8: Delgado used common names for the monkeys—Ali, Lou, Sarah, and Elsa—but I'll stick with alphanumeric designations.

9: José M.R. Delgado. 1963. "Cerebral Heterostimulation in a Monkey Colony." *Science.*

10: M.P. Bishop, S. Thomas Elder, and Robert G. Heath. 1963. "Intracranial Self-Stimulation in Man." *Science.*

11: Robert G. Heath. 1963. "Electrical Self-Stimulation of the Brain in Man." *American Journal of Psychiatry.*

12: Robert G. Heath. 1963. "Electrical Self-Stimulation of the Brain in Man." *American Journal of Psychiatry.*

13: José M.R. Delgado. 1967. "Brain research and behavioural activity." *Endeavour.*

14: José M.R. Delgado. 1969. *Physical Control of the Mind: Toward a Psychocivilized Society.*

15: José M.R. Delgado. 1969. *Physical Control of the Mind: Toward a Psychocivilized Society.*

16: José M.R. Delgado. 1969. *Physical Control of the Mind: Toward a Psychocivilized Society.*

17: A warning, though, for those of a more sensitive nature: the book is filled with pictures of animals with wires and plugs sticking out of their heads.

18: Alan A. Baumeister. 2000. "The Tulane Electrical Brain Stimulation Program: A Historical Case Study in Medical Ethics." *Journal of the History of the Neurosciences.*

19: Eric Halgren and Patrick Chauvel. 1993. "Experiential Phenomena Evoked by Human Brain Electrical Stimulation." In *Advances in Neurology*, volume 63: *Electrical and Magnetic Stimulation of the Brain and Spinal Cord*, edited by Orrin Devinsky, Aleksandar Berić, and Michael Dogali.

20: Alan A. Baumeister. 2000. "The Tulane Electrical Brain Stimulation Program: A Historical Case Study in Medical Ethics." *Journal of the History of the Neurosciences.*

21: I haven't read this book so I withhold judgment. It was printed by Moran Printing, a company located in Baton Rouge, Louisiana.

22: José M.R. Delgado. 1981. "Depth Stimulation of the Brain." *Electrical Stimulation Research Techniques.*

23: This is one of the few statements I've found that acknowledge Delgado's contributions to the development of the modern form of treatment called deep brain stimulation. See chapter 8 for a discussion of this technique.

24: For example, Edward Jenner used a poor young farm boy as a guinea pig to test his smallpox vaccine.

25: See, for example, Dominic Streatfeild's book *Brainwash: The Secret History of Mind Control*.


## Chapter 7


1: The means of transduction is interesting but beyond the scope of this book. It generally involves chemical or mechanical processes along with certain types of ion channels that influence the flow of charged particles through the cell's membrane.

2: Rube Goldberg was an American cartoonist and inventor famous for drawings of overly complex machines, such as an alarm clock in which a ball bearing goes through a tortuous obstacle course and flies off ramps and enters little cylinders all for the ultimate purpose of hitting a bell and waking a sleeping person.

3: Hair cells are simple receptors and lack axons. The cochlear nerve axons come from neurons located within the cochlea. These neurons make synaptic connections with hair cells and transmit the information to the central nervous system. There seems to be slightly more neurons than hair cells so the connections aren't quite one-to-one, but the brain is typically more complicated than a simple one-to-one scheme.

4: Such as loud concerts. Or working on a flight line around noisy planes, which I did as a young airman in the Air Force. The normal hearing range in young humans is 20 to about 20,000 hertz, though people are most responsive to frequencies of around several hundred to several thousand hertz. Due to cell loss at the upper range, adults lose sensitivity off the top end, and older adults max out at perhaps 10,000-15,000 hertz or less (mine is a lot less).

5: Blake S. Wilson and Michael F. Dorman. 2008. "Cochlear implants: Current designs and future possibilities." *Journal of Rehabilitation Research and Development.*

6: Robert Sparrow. 2005. "Defending Deaf Culture: The Case of Cochlear Implants." *The Journal of Political Philosophy.*

7: Daniel J. Felleman and David C. Van Essen. 1991. "Distributed Hierarchical Processing in the Primate Cerebral Cortex." *Cerebral Cortex.*

8: Since rods are predominantly located in the periphery, our peripheral vision is much better in the night time. This means that you'll have better luck seeing a dim object such as a faint star at night by looking slightly to one side of it instead of looking straight at it. If you look right at it then the image falls onto your fovea rather than the rod-rich periphery; at night, it's better to put the image slightly off center, where you've still got a dense packing of photoreceptors but most of them are rods.

9: Half of the axons of each of the two optic nerves cross over to the other hemisphere before they reach the cortex. In other words, about half of the left eye's projection goes to the left hemisphere's cortex and half to the other side, and similarly for the right eye. These projections are arranged so that each cerebral hemisphere receives information from the opposite visual field (the left hemisphere processes the right visual field and the right hemisphere processes the left). This is in keeping with what seems to be the standard operating procedure for the brain: each cerebral hemisphere senses and controls the opposite side of the body. Another interesting feature of the optic nerve is that a few axons project somewhere other than the visual cortex. These axons carry information to other neural networks that need to know such things as the amount of ambient light (for example, neurons that are responsible for your internal clock might use this information to distinguish day time and night time). Such information is also probably what provides some blind patients a peculiar kind of "vision" known as blindsight. The flow of information from the eye to the cortex is disrupted in these patients and so they are blind, but if the other axons are spared then the patients may be able to react subconsciously to light or visual objects. For example, if they are told an object is present, they may be able to reach out with a hand and grab it with uncanny accuracy even though the object doesn't register in their visual consciousness.

10: A receptive field is the region in which the neuron is sensitive to stimuli. For example, a visual neuron may have a receptive field that covers a small amount of space located in some region in space, perhaps just in front of you or a little to the left or right.

11: Wm. H. Dobelle. 2000. "Artificial Vision for the Blind by Connecting a Television Camera to the Visual Cortex." *ASAIO Journal.*

12: The researchers were L. Riggs and F. Ratliff in one group and R.W. Ditchburn and B.L. Ginsborg in the other. Nobel laureate David Hubel describes adaptation and much more in his book *Eye, Brain, and Vision*. This book has recently become available on the Internet. Hubel was a pioneer in the study of the neural basis of vision, and he is as good a writer as he is a researcher. *Eye, Brain, and Vision* is the best book I have read on the visual system.

13: Argus is a multi-eyed creature from Greek my-thology.

14: Duncan Graham-Rowe. 2011. "A Bionic Eye Comes to Market." *Technology Review.*

15: For comparison, the moon subtends an angle of about 1/2 of a degree. An angle of 1/60th of a degree is quite small, approximately the size that a dime would appear at a distance of about 200 feet (61 meters).

16: James D. Weiland and Mark S. Humayun. 2008. "Visual Prosthesis." *Proceedings of the IEEE.*

17: The small eye movements that I referred to earlier in the chapter are known as microsaccades. Microsaccades are involuntary, whereas saccades involve a voluntary change in gaze.

18: Wm. H. Dobelle. 2000. "Artificial Vision for the Blind by Connecting a Television Camera to the Visual Cortex." *ASAIO Journal.*

Chapter 8

1: Or not even present. Amputees often feel pain that curiously seems to emanate from the missing limb. V.S. Ramachandran's book *Phantoms in the Brain* gives an excellent description of this phenomenon, which is called phantom pain.

2: Alon Y. Mogilner, Alim-Louis Benabid, and Ali R. Rezai. 2001. "Brain stimulation: current applications and future prospects." *Thalamus & Related Systems.*

3: The ventricles of the brain are an interconnected network of chambers filled with cerebrospinal fluid. This fluid protects and cushions the brain by providing support from the inside—by filling the ventricles, which are buried within the cerebrum—and the outside (the fluid also flows in the meninges). The ventricles consist of two paired lateral cavities (one for each hemisphere) and a third and fourth cavity along the midline. An aqueduct connects the third and fourth ventricle (and here is where the periacqueductal gray is located).

4: Richard G. Bittar, Ishani Kar-Purkayastha, Sarah L. Owen, Renee E. Bear, Alex Green, ShouYan Wang, and Tipu Z. Aziz. 2005. "Deep brain stimulation for pain relief: A meta-analysis."

*Journal of Clinical Neuroscience.*

5: The substantia nigra got its name because it looks like a black substance in preserved brains (nigra derives from a Latin word meaning black). I remember specifically searching for it in that anatomy class I mentioned in chapter 1. It is one of the easiest nuclei to identify once you've managed to find it—unless the brain belonged to a person with Parkinson's, in which case the nucleus may be nearly gone.

6: The basal ganglia get their name from their position in the brain—lower or basal—and an old fashioned word for nucleus or body, ganglion, the plural of which is ganglia. Not to be confused with a ganglion cell in the retina, though the name of this cell also refers to body—in this case the body of the optic nerve.

7: Sometimes the opinions of scientists, particularly prominent members of the research community, become elevated to the status of hypotheses and theories, even if the scientist has no evidence to back them up. If the scientist is a world-famous Nobel Prize winner, opinions can even pose as facts.

8: All cells have a voltage across their membrane due to the ionic concentrations in the body, but in electrically excitable cells such as neurons and muscle cells, the voltage can change quickly. Normally the voltage inside the cell with respect to ground (neutral) is negative, and is represented as a negative value. For example, a common resting potential of a neuron is -70 millivolts. A neuron will tend to fire an action potential if, say, the potential rises to -55 millivolts or so; because the potential is going toward 0 (getting more positive), neuroscientists call it depolarization, or in other words, decreasing polarization (the potential is getting numerically smaller as it heads toward 0). During an action potential, a neuron's voltage briefly rises to a level generally slightly above 0 volts, then drops back down to below its resting level before returning to its normal state.

9: Neuroscientists call this hyperpolarization because it makes the neuron's potential more negative inside, farther away from zero—in other words, more (hyper) polarized.

10: Animal models of Parkinson's disease often use a toxic drug known as MPTP (which is an abbreviation for an onerously long chemical name). The drug kills neurons in the substantia nigra. It was discovered when some drug abusers took an illicit drug tainted with MPTP and came down with Parkinson's disease.

11: Many of these studies are summarized in review papers published in science journals. These information-dense papers cite a lot of other publications and often provide a brief commentary on the findings and how they fit into or fail to fit into prevailing wisdom. They don't make light reading because they're written for other researchers, but if you're doggedly interested in pursuing the subject, recent reviews on the physiological effects of DBS include the following:

Cameron C. McIntyre, Marc Savasta, Lydia Kerkerian-Le Goff, and Jerrold L. Vitek. 2004. "Uncovering the mechanism(s) of action of deep brain stimulation: activation, inhibition, or both." *Clinical Neurophysiology.*

Joel S. Perlmutter and Jonathan W. Mink. 2006. "Deep Brain Stimulation." *Annual Review of Neuroscience.*

Erwin B. Montgomery Jr. and John T. Gale. 2007. "Mechanisms of action of deep brain stimulation (DBS)." *Neuroscience and Biobehavioral Reviews.*

Jean-Michel Deniau, Bertrand Degos, Clémentine Bosch, and Nicolas Maurice. 2010. "Deep brain stimulation mechanisms: beyond the concept of local functional inhibition." *European Journal of Neuroscience.*

12: Benjamin D. Greenberg and Ali R. Rezai. 2003. "Mechanisms and the Current State of Deep Brain Stimulation in Neuropsychiatry." *CNS Spectrums.*

13: Animal models of depression have varying degrees of plausibility. A model isn't the actual disease but a possible representation or simulation of the disease. The model abstracts what researchers believe are the essential aspects of the thing being modeled. If the model is wrong, all of the research based upon it may have little applicability.

14: Depression is a symptom of a number of other diseases. One such disease is bipolar disorder (sometimes still called by its old name, manic depression), in which the patient's mood alternates between extremely high and extremely low states. Bipolar patients aren't generally treated with the same therapies as those who exhibit depression without mania. To highlight the distinction between the two diseases, depression is sometimes called unipolar depression. But psychiatrists can confuse bipolar disease with unipolar depression if the manic phases don't last long or appear infrequently.

15: Donald A. Malone Jr. 2010. "Use of deep brain stimulation in treatment-resistant depression." *Cleveland Clinic Journal of Medicine.*

16: B.H. Bewernick et al. 2010. "Nucleus accumbens deep brain stimulation decreases ratings of depression and anxiety in treatment-resistant depression." *Biological Psychiatry.*

17: Bettina H. Bewernick, Sarah Kayser, Volker Sturm, and Thomas E. Schlaepfer. 2012. "Long-Term Effects of Nucleus Accumbens Deep Brain Stimulation in Treatment-Resistant Depression: Evidence for Sustained Efficacy." *Neuropsychopharmacology.*

18: Korbinian Brodmann painstakingly conducted a microscopic examination of all the tissue of the cerebral cortex in man and monkey. He parceled the cortex into areas identified by distinct cellular structure—types of cells and their arrangement in the tissue. This formed a map of the surface of the brain and was often used, along with other similar maps, to give a name (actually, a number) to specific regions of the cortex. Brodmann's map had much greater precision than just naming extensive regions of cortex based on the location of the bones of the skull, as in occipital, parietal, temporal and frontal cortex. Note that Brodmann did *not* make a functional map of the cortex, but rather one based on differences in histology (tissue). As neuroscientists began recording the electrical activity of the cortex, they began labeling cortical regions by their function—primary visual cortex, auditory cortex, and so on. In some cases Brodmann's areas coincide quite well with functionality; for example, area 17 corresponds with primary visual cortex. But these days researchers tend to dispense with Brodmann's map and use functional names, although sometimes when an area has no known specific function, researchers go back to Brodmann.

19: Yet another naming scheme, based on the brain's furrowed geography. Subcallosal means it's beneath the corpus callosum, a thick band of fibers connecting the two hemispheres. Cingulate is derived from a Latin word meaning belt or girdle, and refers to the way that this gyrus borders or girdles the corpus callosum.

20: Emory University press release. 9 January 2012. "Deep Brain Stimulation Shows Promising Results For Unipolar and Bipolar Depression."

21: *Medical News Today*. 2007. "Deep Brain Stimulation Helps Near Vegetative Man Say 'I Love You Mommy'."

22: ClinicalTrials.gov identifier NCT01512134. The address for the web page is: http://clinicaltrials.gov/ct2/show/NCT01512134.

23: William P. Melega, Goran Lacan, Alessandra A. Gorgulho, Eric J. Behnke, and Antonio A.F. De Salles. 2012. "Hypothalamic Deep Brain Stimulation Reduces Weight Gain in an Obesity-Animal Model." PLoS ONE.

## Chapter 9

1: By "changing" I simply mean that the magnetic field varies over time. It can change in any kind of way. A steady magnetic field has no effect on a charge unless the charge is already moving, in which case the force is perpendicular to the charge's path. This might seem weird but that's just the way it is.

2: Fernand Papillon. 1873. "Electricity and Life." *The Popular Science Monthly*. This article was originally published in the French journal *Revue des Deux Mondes*. In those days magazines legally and routinely lifted articles appearing in other countries and published them, suitably translated, in their own pages. At least *Popular Science Monthly* credited the source—some of the other magazines didn't bother.

3: William J. Herdman. 1901. "The Necessity for Special Education in Electro-Therapeutics." In *An International System of Electro-Therapeutics*, edited by Horatio R. Bigelow and G. Betton Massey, 2nd edition.

4: The American Medical Association (AMA) web site, located at: http://www.ama-assn.org.

5: L.A. Geddes. 1984. "The Beginnings of Electromedicine." *IEEE Engineering in Medicine and Biology Magazine*.

6: Not much fat can pass through the blood-brain barrier and the brain doesn't have its own storage facility. During a long period of starvation the liver produces substances called ketones that are derived from fatty acids and can cross the blood-brain barrier. Ketones thus supply energy in times of crisis, but an insulin coma can occur before they have a chance to get revved up.

7: It is interesting how much influence Hollywood seems to have, even though movie writers and producers rarely achieve any sort of historical or scientific accuracy, and seldom do they even appear to strive for it. Another interesting note about this movie is that, like many influential movies, the story was originally a novel, written by Ken Kesey and published in 1962. It's ironic that a movie based on a book by Kesey should play a significant role in besmirching a peculiar therapy. In the 1960s Kesey was a famously outspoken advocate of LSD "therapy."

8: Jennifer S. Perrin, Susanne Merz, Daniel M. Bennett, James Currie, Douglas J. Steele, Ian C. Reid, and Christian Schwarzbauer. 2012. "Electroconvulsive therapy reduces frontal cortical connectivity in severe depressive disorder." *Proceedings of the National Academy of Sciences.*

9: See, for example—if you can find a copy—a 1970 book by Sheila Ostrander and Lynn Schroeder, *Psychic Discoveries Behind the Iron Curtain.*

10: F. Rubin. 1966. "Electric Sleep Therapy." *New Scientist.*

11: Several theories have emerged to explain the functions of sleep. Some of the more interesting ones include a role for sleep in memory formation and wound healing. Whatever its functions, sleep appears fundamental. All mammals sleep, though sea mammals apparently do it differently—only half of the brain shuts down at any given time—and some biologists say it's not really "sleep."

12: Wallace B. Pickworth, Reginald V. Fant, Marsha F. Butschky, Allison L. Goffman, and Jack E. Henningfield. 1997. "Evaluation of Cranial Electrostimulation Therapy on Short-Term Smoking Cessation." *Biological Psychiatry.*

13: Harald Walach and Eduard Käseberg. 1998. "Mind Machines: A Controlled Study on the Effects of Electromagnetic and Optic-Acoustic Stimulation on General Well-Being, Electrodermal Activity, and Exceptional Psychological Experiences." *Behavioral Medicine.*

14: Electromagnetic radiation possesses the bizarre ability to behave as waves *and* particles, though not at the same time. The particles are known as photons. This peculiar phenomenon, known as wave-particle duality, has not been fully explained.

15: To visualize the inverse square law, imagine a pulse of radiation emitted from a point source. The radiation spreads in all directions, creating a spherical wave front. As the sphere expands its surface area increases, but the energy it contains is fixed at the point of departure. This means that the energy density must decrease. (Think of a balloon—as it blows up the fabric must stretch.) The surface area of a sphere is proportional to the square of its radius, and if you divide a given amount of energy by this area you get the inverse square law.

16: Herbert Pollack. 1979. "Epidemiologic Data on American Personnel in the Moscow Embassy." *Bulletin of the New York Academy of Medicine.*

17: Kathleen McAuliffe. 1985. "The Mind Fields." *Omni.*

18: Some animals such as certain birds and lobsters can detect magnetic fields, a sense called magnetoception. These animals appear to use Earth's magnetic field for orientation and navigation. Every once in a while I run across stories claiming that people also have a magnetic sense, but that's never been proven.

19: Laura Helmuth. 2001. "Boosting Brain Activity From The Outside In." *Science.*

20: Alvaro Pascual-Leone, Vincent Walsh, and John Rothwell. 2000. "Transcranial magnetic stimulation in cognitive neuroscience—virtual lesion, chronometry, and functional connectivity." *Current Opinion in Neurobiology.*

21: Carlo Miniussi, Manuela Ruzzoli, and Vincent Walsh. 2010. "The mechanism of transcranial magnetic stimulation in cognition." *Cortex.*

22: As reviewed in: Soroush Zaghi, Mariana Acar, Brittney Hultgren, Paulo S. Boggio, and Felipe Fregni. 2009. "Noninvasive Brain Stimulation with Low-Intensity Electrical Currents: Putative Mechanisms of Action for Direct and Alternating Current Stimulation." *Neuroscientist.*

23: Jerome Brunelin et al. 2012. "Examining Transcranial Direct-Current Stimulation (tDCS) as a Treatment for Hallucinations in Schizophrenia." *American Journal of Psychiatry.*

24: The prefrontal cortex is often associated with the "highest" cognitive functions since it seems to be active while people are engaged in complex decision-making or problem solving. As the name suggests, it is located in the most anterior (frontal) part of the brain. The prefrontal region known as the dorsolateral prefrontal cortex seems to be particularly important.

25: Colleen K. Loo, Angelo Alonzo, Donel Martin, Philip B. Mitchell, Veronica Galvez, and Perminder Sachdev. 2012. "Transcranial direct current stimulation for depression: 3-week, randomised, sham-controlled trial." *British Journal of Psychiatry.*

Chapter 10

1: For example, Gary Marcus's 2008 book *Kluge: The Haphazard Construction of the Human Mind.*

2: For example, Russian psychologist A.R. Luria spent years studying a person he identified as "S" who seemed unable to forget anything. The story is recounted in Luria's book *The Mind of a Mnemonist.*

3: Felipe Fregni et al. 2005. "Anodal transcranial direct current stimulation of prefrontal cortex enhances working memory." *Experimental Brain Research.*

4: Suk Hoon Ohn et al. 2008. "Time-dependent effect of transcranial direct current stimulation on the enhancement of working memory." *NeuroReport.*

5: Felipe Fregni et al. 2008. "Transcranial direct current stimulation of the prefrontal cortex modulates the desire for specific foods." *Appetite.*

6: Allan W. Snyder, Elaine Mulcahy, Janet L. Taylor, D. John Mitchell, Perminder Sachdev, and Simon C. Gandevia. 2003. "Savant-like skills exposed in normal people by suppressing the left fronto-temporal lobe." *Journal of Integrative Neuroscience.*

7: ALS (also known as Lou Gehrig's disease) attacks neurons in the brain that mediate voluntary movement.

8: Simultaneous recording of the activity of individual neurons is an entirely different technique than something like electroencephalography. An EEG displays the simultaneous activity of many neurons but the contributions of these neurons are smeared together so that you can't identify any individual cell.

9: John K. Chapin, Karen A. Moxon, Ronald S. Markowitz, and Miguel A.L. Nicolelis. 1999. "Real-time control of a robot arm using simultaneously recorded neurons in the motor cortex." *Nature Neuroscience.*

10: Johan Wessberg et al. 2000. "Real-time prediction of hand trajectory by ensembles of cortical neurons in primates." *Nature.*

11: Cyborg is a portmanteau word coined in 1960 from the term *cybernetic organism.* Cybernetic—or cybernetics, as it is most common used today—is a term from the 1940s that has yet to settle down to a common definition, though dictionaries try to capture the essence by defining it as the science of communication and control. (It derives from a Greek word meaning to govern.) In the 1940s researchers began to design machines capable of operating or navigating on their own, which generally required automated systems to provide feedback, detect errors, and make corrective adjustments. Scientists often used "cybernetics" in the description of these devices. The modern use of the term *cyborg* goes well beyond the sense of automation to include the use of machines and biological systems together. Some people also use the term *bionics* to describe the marriage of man and machine.

12: Joseph E. O'Doherty, Mikhail A. Lebedev, Timothy L. Hanson, Nathan A. Fitzsimmons, and Miguel A.L. Nicolelis. 2009. "A brain-machine interface instructed by direct intracortical micro-stimulation." *Frontiers in Integrative Neuroscience.*

13: These sophisticated statistical calculations are invariably done entirely by computer and are only correct if the researchers used the correct software package—something I fear doesn't happen very often, and even when it does it is a purely fortunate coincidence. Most neuroscientists don't understand statistics and were never adequately educated in the subject.

14: P.R. Kennedy, and R.A.E. Bakay. 1998. "Restoration of neural output from a paralyzed patient by a direct brain connection." *NeuroReport.*

15: Sanjiv K. Talwar, Shaohua Xu, Emerson S. Hawley, Shennan A. Weiss, Karen A. Moxon, and John K. Chapin. 2002. "Rat navigation guided by remote control." *Nature.*

16: Miguel A.L. Nicolelis. 2002. "The amazing adventures of robotrat." *Trends in Cognitive Sciences.*

17: C. Daniel Salzman, Kenneth H. Britten, and William T. Newsome. 1990. "Cortical microstimulation influences perceptual judgements of motion direction." *Nature.*

18: MT is also known as V5, part of the chain of visual areas starting with primary visual cortex (which is also known as V1).

19: Seyed-Reza Afraz, Roozbeh Kiani, and Hossein Esteky. 2006. "Microstimulation of inferotemporal cortex influences face categorization." *Nature.*

20: Itzhak Fried, Charles L. Wilson, Katherine A. MacDonald, Eric J. Behnke. 1998. "Electric current stimulates laughter." *Nature.*

Chapter 11

1: Alberto Priori, et al. 2008. "Lie-Specific Involvement of Dorsolateral Prefrontal Cortex in Deception." *Cerebral Cortex.*

2: Ahmed A. Karim, et al. 2010. "The Truth about Lying: Inhibition of the Anterior Prefrontal Cortex Improves Deceptive Behavior." *Cerebral Cortex.*

3: Statistics are evaluated in terms of the probability that the result of the experiment could be due to chance. The researcher chooses this probability, often 0.05, or 1 in 20. What this means is that there is only a 1 in 20 chance that the results could be due to some chance fluctuation rather than a real effect. Some researchers choose an even lower probability, say 1 in 100 (0.01), just to make sure. Results that exceed this level of probability are called statistical significant. If they fail to reach this level, they are—by strict interpretation of the assumptions, and indeed the mathematical foundations of statistics—considered insignificant. So even if your results seem to show a certain effect, statistics may indicate that it could easily be due to chance, and so it should be disregarded. This can be quite disheartening.

4: Inga Karton and Talis Bachmann. 2011. "Effect of prefrontal transcranial magnetic stimulation on spontaneous truth-telling." *Behavioural Brain Research.*

5: The 4th amendment protects Americans from unreasonable searches, but what exactly is "unreasonable"? In my view, it's unreasonable for law enforcement officers to search anyone unless the officers have good cause to suspect the person of engaging in illegal activity. Yet the Transportation Security Administration searches passengers solely because they are about to board an airplane. Is that good cause? Although I can't cite any statistics, I'd be willing to bet that 99.999 percent of people who board an airplane are planning on arriving safely at their destination. If the act of boarding a plane is sufficient reason for the government to conduct a search, then almost anything else is also sufficient.

6: I can imagine a totalitarian government that could force people to undergo brain stimulation. But if we lived in a society that tolerated such a government, we'd have bigger problems than just unwanted brain stimulation.

7: Not that I'm advocating sneakiness. What I'm doing here is exploring the potential of brain stimulation. The nature of its application and whether it will be welcome or not is another subject.

8: These molecules can form a "cage" that binds other molecules and releases them only when a light shines on the ensemble. This process can be used to deliver small molecules in an inactive state to a specific location in the body and then activate them once they arrive at the target. Side effects can be reduced because the molecules do not interact with tissues other than the target.

9: This part of the spectrum is called near infrared because it is in the infrared range—slightly lower in frequency than red light, which has the lowest frequency of the visible spectrum—but it is a slice of infrared that is close to red. Far infrared has frequencies closer to microwaves, which occupy the next lowest frequency rung of the spectrum.

10: Some recent papers on optogenetics from Karl Deisseroth and his colleagues include the following:

Karl Deisseroth. 2010. "Controlling the Brain with Light." *Scientific American.*

Ofer Yizhar, Lief E. Fenno, Thomas J. Davidson, Murtaza Mogri, and Karl Deisseroth. 2011. "Optogenetics in Neural Systems." *Neuron.*

Lief Fenno, Ofer Yizhar, and Karl Deisseroth. 2011. "The Development and Application of Optogenetics." *Annual Review of Neuroscience.*

Polina Anikeeva et al. 2012. "Optetrode: a multichannel readout for optogenetic control in freely moving mice." *Nature Neuroscience.*

11: Anita Mahadevan-Jansen et al. 2010. "Imaging optically induced neural activity in the brain." *Conference Proceedings of the*

*IEEE Engineering in Medicine and Biology Society.*

12: Some people also call dopamine as well as other molecules "happy chemicals" for similar reasons. People who use this term tend to be people who have learned all they know about neuroscience from television talk shows.

# Glossary

**AC** alternating current (see entry)

**action potential** a brief spike in voltage occurring in a neuron, and which travels along the neuron's axon and carries information

**alternating current** an electric current that periodically reverses polarity (direction of flow)

**amygdala** a nucleus (actually a set of nuclei) located in the temporal lobe that plays a role in the processing and regulation of emotion

**axon** a long, thin projection of a neuron that carries information in the form a train of action potentials to signal other neurons

**basal ganglia** group of nuclei located below the cortex that are involved in movement and motor functions

**blood-brain barrier** a filtering mechanism that prevents bulky and potential hazardous substances from seeping through the capillaries and entering the brain

**caudate nucleus** a prominent member of the basal ganglia

**cerebral cortex** a thin, wrinkled layer (set of layers, actually) that covers the cerebral hemispheres, and consists of neural networks that are involved in sensation, perception, movement, and high-level planning and decision-making functions

**cerebral hemisphere** one of two main symmetrical halves of the brain, covered by cerebral cortex

**cortex** in this book, the cerebral cortex (see entry)

**craniotomy** a procedure that creates an opening or hole in the cranium

**DC** direct current (see entry)

**dendrite** a projection of the neuron that generally receives signals from other neurons

**direct current** an electric current that flows in only one direction, such as the current produced by a battery

**DLPFC** dorsolateral prefrontal cortex (see entry)

**DNA** deoxyribonucleic acid, the material that comprises the double helix and contains genes

**dorsolateral prefrontal cortex** an important region in the front of the brain that is involved in high-level cognitive functions such as decision-making and planning

**EEG** electroencephalogram (or electroencephalography), the recording (or process of recording) electrical signals from the brain

**electrode** a current-carrying conductor, which can be a small wire inserted into the brain or a bead or plate that is attached to the scalp

**fMRI** functional magnetic resonance imaging (see entry)

**frontal cortex** the anterior part of the cortex, many parts of which are involved in higher level cognitive functions

**frontal lobe** the anterior section of the cerebral hemisphere

**functional magnetic resonance imaging** a method of measuring relative activity in the brain based on blood flows and metabolism

**gene** a unit of heredity that is physically represented by a stretch of DNA and codes for a specific protein or ribonucleic acid (RNA) sequence

**gene expression** the activation of a gene and the mechanism by which its product gets made

**glia (glial cell)** a class of cell types in the brain mostly devoted to maintenance or structural functions

**gyrus (plural: gyri)** a ridge in the cortex, surrounded by valleys (sulci)

**hemisphere** with respect to the brain, a cerebral hemisphere (see entry)

**hypothalamus** a small but critical group of related nuclei at the base of the brain that are involved in the regulation of certain fundamental hormones and behaviors such as eating, drinking, and aggression

**in vitro** occurring on a plate or dish in the laboratory

**in vivo** occurring in the body

**ion** charged particle

**ion channel** a protein embedded in a cell's membrane that permits the passage of ions (electric charges generally can't pass through the membrane itself)

**membrane** a thin layer made mostly of lipids (fats) that surround and protect a cell's interior

**meninges**  three layers (often called membranes, though they are not the same thing as cellular membranes) of tissue that surround and cushion the brain and spinal cord

**metabolism**  a general term for the set of biochemical reactions occurring in a cell, tissue, or organ

**microstimulation**  stimulation with a small ("micro") electrode

**motor cortex**  a strip of cortex involved in the control of voluntary movements

**nerve**  a bundle of axons that carry signals to and from the brain

**neural network**  a circuit of synaptically coupled neurons that perform a certain function

**neuron**  an electrically active cell in the brain involved in the processing of information

**neurotransmitter**  a signaling chemical that is released at points (terminals) of axons and diffuses to the membrane of other neurons, usually across gaps known as synapses

**parietal cortex**  cortex that covers the parietal lobe and performs a variety of functions related to sensations in the body and movement or orientation of the body in space

**parietal lobe**  section of the cerebral hemisphere at the top or dome of the head

**PET**  positron emission tomography (see entry)

**placebo effect**  an improvement or change in condition that is not directly attributable to any medical treatment

**positron emission tomography** an imaging technique that measures metabolic activity by detecting the emissions of radioactive molecules

**prefrontal cortex** region of the cortex lying just behind the forehead that is involved in "executive" functions

**prosthesis** an artificial device that replaces or supplements a missing or impaired part of the body

**protein** functional molecule made of amino acids, whose sequence is specified by a gene

**rTMS** repetitive form of transcranial magnetic stimulation (see entry)

**septal area (region)** a group of nuclei located deep in the brain that are involved in certain aspects of emotion and reinforcement

**somatosensory cortex** strip of cortex in the anterior part of the parietal lobe that detects touch or pressure sensations of the body

**subcortical nucleus** a collection of cells located below the cortex

**sulcus** a valley or furrow in the cortex

**synapse** a junction or gap between two neurons, usually consisting of a terminal of one neuron's axon and the membrane (often part of a dendrite) of another neuron; chemicals known as neurotransmitters are released from the axon terminal and activate receptors in the other neuron as part of the signaling process

**tDCS** transcranial direct current stimulation (see entry)

**tegmentum** a region of cells and axons located at the base of the brain which is involved in regulation of basic physiological processes such as awareness and movement

**temporal cortex** area of cortex covering the temporal lobe, and involved in variety of functions such as audition, language, and memory

**temporal lobe** region of the cerebral hemisphere at the side of the head

**thalamus** collection of nuclei deep in the brain, many of which are extensively connected with the cortex and perform a wide variety of functions related to sensation and movement

**TMS** transcranial magnetic stimulation (see entry)

**transcranial direct current stimulation** the stimulation of the brain with electrodes attached to the skull, which sends current through the (intact) cranium

**transcranial magnetic stimulation** the use of strong magnets to generate a pulse (or a train of pulses, if repetitive stimulation is employed) that stimulates the brain by creating electric fields

**ventricle** with respect to the brain, one of a set of fluid-filled cavities within the brain that serves as a cushion and also carries away waste products

## About the Author

Kyle Kirkland earned a Ph.D. in neuroscience and has worked on models of the brain as well as brain stimulation experiments. He is now a writer of science fact and fiction.

Printed in Great Britain
by Amazon